Mathematics of Harmony as a New Interdisciplinary Direction and "Golden" Paradigm of Modern Science

Volume 3

The "Golden" Paradigm of Modern Science:
Prerequisite for the "Golden" Revolution in Mathematics,
Computer Science, and Theoretical Natural Sciences

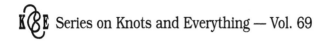
K&E Series on Knots and Everything — Vol. 69

Mathematics of Harmony as a New Interdisciplinary Direction and "Golden" Paradigm of Modern Science

Volume 3
The "Golden" Paradigm of Modern Science:
Prerequisite for the "Golden" Revolution in Mathematics,
Computer Science, and Theoretical Natural Sciences

Alexey Stakhov
International Club of the Golden Section, Canada & Academy of Trinitarism, Russia

World Scientific

EW JERSEY · LONDON · SINGAPORE · BEIJING · SHANGHAI · HONG KONG · TAIPEI · CHENNAI · TOKYO

Published by

World Scientific Publishing Co. Pte. Ltd.

5 Toh Tuck Link, Singapore 596224

USA office: 27 Warren Street, Suite 401-402, Hackensack, NJ 07601

UK office: 57 Shelton Street, Covent Garden, London WC2H 9HE

Library of Congress Cataloging-in-Publication Data

Names: Stakhov, A. P. (Alekseĭ Petrovich), author.

Title: Mathematics of harmony as a new interdisciplinary direction and
    "golden" paradigm of modern science / Alexey Stakhov.

Description: Hackensack, New Jersey : World Scientific, [2020] | Series: Series on knots and
    everything, 0219-9769 ; vol. 69 | Includes bibliographical references. | Contents:
    Volume 1. The golden section, Fibonacci numbers, Pascal triangle, and Platonic solids --
    Volume 2. Algorithmic measurement theory, Fibonacci and golden arithmetic and
    ternary mirror-symmetrical arithmetic -- Volume 3. The "golden" paradigm of modern science.

Identifiers: LCCN 2020010957 | ISBN 9789811207105 (v. 1 ; hardcover) |
    ISBN 9789811213465 (v. 2 ; hardcover) | ISBN 9789811206375 (v. 1 ; ebook) |
    ISBN 9789811213496 (v. 3 ; hardcover) | ISBN 9789811206382 (v. 1 ; ebook other) |
    ISBN 9789811213472 (v. 2 ; ebook) | ISBN 9789811213489 (v. 2 ; ebook other) |
    ISBN 9789811213502 (v. 3 ; ebook) | ISBN 9789811213519 (v. 3 ; ebook other)

Subjects: LCSH: Fibonacci numbers. | Golden section. | Mathematics--History. | Science--Mathematics.

Classification: LCC QA246.5 .S732 2020 | DDC 512.7/2--dc23

LC record available at https://lccn.loc.gov/2020010957

British Library Cataloguing-in-Publication Data

A catalogue record for this book is available from the British Library.

For any available supplementary material, please visit
https://www.worldscientific.com/worldscibooks/10.1142/11645#t=suppl

Desk Editor: Liu Yumeng

Typeset by Stallion Press
Email: enquiries@stallionpress.com

*In fond memory of Yuri Alekseevich Mitropolskiy
and
Alexander Andreevich Volkov*

Contents

Preface to the Three-Volume Book

Continuity in the Development of Science

Scientific and technological progress has a long history and passed in its historical development several stages: The Babylonian and Ancient Egyptian culture, the culture of Ancient China and Ancient India, the Ancient Greek culture, the Middle Ages, the Renaissance, the Industrial Revolution of the 18th century, the Great Scientific Discoveries of the 19th century, the Scientific and Technological Revolution of the 20th century and finally the 21st century, which opens a new era in the history of mankind, the *Era of Harmony*.

Although each of the mentioned stages has its own specifics, at the same time, every stage necessarily includes the content of the preceding stages. This is called the *continuity* in the development of science.

It was during the ancient period, a number of the fundamental discoveries in mathematics were made. They exerted a determining influence on the development of the material and spiritual culture. We do not always realize their importance in the development of mathematics, science, and education. To the category of such discoveries, first of all, we must attribute the *Babylonian numeral system with the base* 60 and the *Babylonian positional principle of number representation*, which is the foundation of the, *decimal*, *binary*, *ternary*, and other positional numeral systems. We must add to this list the *trigonometry* and the *Euclidean geometry*, the *incommensurable segments* and the *theory of irrationality*, the *golden section* and *Platonic solids*, the *elementary number theory* and the *mathematical theory of measurement*, and so on.

The *continuity* can be realized in various forms. One of the essential forms of its expression are the fundamental scientific ideas, which permeate all stages of the scientific and technological progress and influence various areas of science, art, philosophy, and technology. The idea of *Harmony*, connected with the *golden section*, belongs to the category of such fundamental ideas.

According to B.G. Kuznetsov, the researcher of Albert Einstein's creativity, the great physicist piously believed that science, physics in particular, always had its eternal fundamental goal *"to find in the labyrinth of the observed facts the objective harmony"*. The deep faith of the outstanding physicist in the existence of the universal laws of the *Harmony* is evidenced by another well-known Einstein's statement: *"The religiousness of the scientist consists in the enthusiastic admiration for the laws of the Harmony"* (the quote is taken from the book *Meta-language of Living Nature* [1], written by the outstanding Russian architect Joseph Shevelev, known for his research in the field of *Harmony* and the *golden section* [1–3]).

Pythagoreanism and Pythagorean MATHEM's

By studying the sources of the origin of mathematics, we inevitably come to Pythagoras and his doctrine, named the *Pythagoreanism* (see Wikipedia article *Pythagoreanism*, the Free Encyclopedia). As mentioned in Wikipedia, the *Pythagoreanism* originated in the 6th century BC and was based on teachings and beliefs of Pythagoras and his followers called the Pythagoreans. Pythagoras established the first Pythagorean community in Croton, Italy. The Early Pythagoreans espoused a rigorous life and strict rules on diet, clothing and behavior.

According to tradition, *Pythagoreans* were divided into two separate schools of thought: the *mathematikoi* (*mathematicians*) and the *akousmatikoi* (*listeners*). The *listeners* had developed the religious and ritual aspects of *Pythagoreanism*; the *mathematicians* studied the four Pythagorean MATHEMs: *arithmetic*, *geometry, spherics*, and *harmonics*. These MATHEMs, according to Pythagoras, were the main composite parts of mathematics. Unfortunately, the Pythagorean MATHEM of the *harmonics* was lost in mathematics during the process of its historical development.

Proclus Hypothesis

The Greek philosopher and mathematician Proclus Diadoch (412–485 AD) put forth the following unusual hypothesis concerning Euclid's *Elements*. Among Proclus's mathematical works, his *Commentary on the Book I of Euclid's Elements* was the most well known. In the commentary, he puts forth the following unusual hypothesis.

It is well known that Euclid's *Elements* consists of 13 books. In those, XIIIth book, that is, the concluding book of the *Elements*, was devoted to the description of the geometric theory of the five *regular polyhedra*, which had played a dominant role in *Plato's cosmology* and is known in modern science under the name of the *Platonic solids*.

Proclus drew special attention to the fact that the concluding book of the *Elements* had been devoted to the *Platonic solids*. Usually, the most important material, of the scientific work is placed in its final part. Therefore, by placing *Platonic solids* in Book XIII, that is, in the concluding book of his *Elements*, Euclid clearly pointed out on main purpose of writing his *Elements*. As the prominent Belarusian philosopher Edward Soroko points out in [4], according to Proclus, Euclid *"had created his Elements allegedly not for the purpose of describing geometry as such, but with purpose to give the complete systematized theory of constructing the five Platonic solids; in the same time Euclid described here some latest achievements of mathematics"*.

It is for the solution of this problem (first of all, for the creation of geometric theory of *dodecahedron*), Euclid already in Book II introduces Proposition II.11, where he describes the *task of dividing the segment in the extreme and mean ratio* (the *golden section*), which then occurs in other books of the *Elements*, in particular in the concluding book (XIII Book).

But the *Platonic solids* in *Plato's cosmology* expressed the *Universal Harmony* which was the main goal of the ancient Greeks science. With such consideration of the *Proclus hypothesis*, we come to the surprising conclusion, which is unexpected for many historians of mathematics. According to the *Proclus hypothesis*, it turns out that from Euclid's *Elements*, two branches of mathematical sciences had originated: the **Classical Mathematics**, which included the *Elements* of the *axiomatic approach* (Euclidean axioms), *the elementary number theory*, and *the theory of irrationalities*, and the Mathematics of Harmony, which was based on the geometric *"task of dividing the segment in the extreme and mean ratio"* (the *golden section*) and also on the theory of the *Platonic solids*, described by Euclid in the concluding Book XIII of his *Elements*.

The Statements by Alexey Losev and Johannes Kepler

What was the main idea behind ancient Greek science? Most researchers give the following answer to this question: **The idea of Harmony connected to the *golden section*.** As it is known, in ancient Greek philosophy, *Harmony* was in opposition to the *Chaos* and meant the organization of the Universe, the Cosmos. The outstanding Russian philosopher Alexey Losev (1893–1988), the researcher in the aesthetics of the antiquity and the Renaissance, assesses the main achievements of the ancient Greeks in this field as follows [5]:

> *"From Plato's point of view, and in general in the terms of the entire ancient cosmology, the Universe was determined as the certain proportional whole, which obeys to the law of the harmonic division, the golden section ... The ancient Greek system of the cosmic proportion in the literature is often interpreted as the curious result of the unrestrained and wild imagination. In such explanation we see the scientific helplessness of those, who claim this. However, we can understand this historical and aesthetic phenomenon only in the connection with the holistic understanding of history, that is, by using the dialectical view on the culture and by searching for the answer in the peculiarities of the ancient social life."*

Here, Losev formulates the *"golden"* *paradigm* of ancient cosmology. This paradigm was based upon the fundamental ideas of ancient science that are sometimes treated in modern science as the "*curious result of the unrestrained and wild imagination*". First of all, we are talking about the *Pythagorean Doctrine of the Numerical Universal Harmony* and *Plato's Cosmology* based on the *Platonic solids*. By referring to the geometrical structure of the Cosmos and its mathematical relations, which express the Cosmic Harmony, the Pythagoreans had anticipated the modern mathematical basis of the natural sciences, which began to develop rapidly in the 20th century. Pythagoras's and Plato's ideas about the Cosmic Harmony proved to be immortal.

Thus, the idea of Harmony, which underlies the ancient Greek doctrine of Nature, was the main "paradigm" of the Greek science, starting from Pythagoras and ending with Euclid. This paradigm relates directly to the *golden section* and the *Platonic solids*, which are the most important Greek geometric discoveries for the expression of the Universal Harmony.

Johannes Kepler (1571–1630), the prominent astronomer and the author of "Kepler's laws", expressed his admiration with the *golden ratio* in the following words [6]:

> "*Geometry has the two great treasures: the first of them is the theorem of Pythagoras; the second one is the division of the line in the extreme and mean ratio. The first one we may compare to the measure of the gold; the second one we may name the precious stone.*"

We should recall again that the ancient *task of dividing line segment in extreme and mean ratio* is Euclidean language for the *golden section*!

The enormous interest in this problem in modern science is confirmed by the rather impressive and far from the complete list of books and articles on this subject, published in the second half of the 20th century and the beginning of the 21st century [1–100].

Ancient Greeks Mathematical Doctrine of Nature

According to the outstanding American historian of mathematics, Morris Kline [101], the main contribution of the ancient Greeks is the one "*which had the decisive influence on the entire subsequent culture, was that they took up the study of the laws of Nature"*. The main conclusion, from Morris Kline's book [101] is the fact that the ancient Greeks proposed the innovative concept of the Cosmos, in which everything was subordinated to the mathematical laws. Then the following question arises: during which time this concept was developed? The answer to this question is also addressed in Ref. [101].

According to Kline [101], the innovative concept of the Cosmos based on the mathematical laws, was developed by the ancient Greeks in the period from VI to III centuries BC. But according to the prominent Russian mathematician academician A.N. Kolmogorov [102], in the same period in ancient Greece, *"the mathematics was created as the independent science with the clear understanding of the uniqueness of its method and with the need for the systematic presentation of its basic concepts and proposals in the fairly general form."* But then, the following important question, concerning the history of the original mathematics arises: was there any relationship between the process of creating the mathematical theory of Nature, which was considered as the goal and the main achievement of ancient Greek science [101], and the process of creating mathematics, which happened in ancient Greece in the same period [102]? It turns out that such connection, of course, existed. Furthermore, it can be argued that these processes actually coincided, that is, the processes of the creation of mathematics by the ancient Greeks [102], and their doctrine of Nature, based on the mathematical principles [101], were one and the same processes. And the most vivid embodiment of the process of the *Mathematization of Harmony* [68] happened in Euclid's *Elements*, which was written in the third century BC.

Introduction of the Term *Mathematics of Harmony*

In the late 20th century, to denote the mathematical doctrine of Nature, created by the ancient Greeks, the term *Mathematics of Harmony* was introduced. It should be noted that this term was chosen very successfully because it reflected the main idea of the ancient Greek science, the *Harmonization of Mathematics* [68]. For the first time, this term was introduced in the small article "Harmony of spheres", placed in *The Oxford Dictionary of Philosophy* [103]. In this article, the concept of *Mathematics of Harmony* was associated with the *Harmony of spheres*, which was, also called in Latin as *"harmonica mundi"* or *"musica mundana"* [10]. The *Harmony of spheres* is the ancient and medieval doctrine on the musical and mathematical structure of the Cosmos, which goes back to the Pythagorean and Platonic philosophical traditions.

Another mention about the *Mathematics of Harmony* in the connection to the ancient Greek mathematics is found in the book by Vladimir Dimitrov, *A New Kind of Social Science*, published in 2005 [44]. It is important to emphasize that in Ref. [44], the concept of *Mathematics of Harmony* is directly associated with the *golden section*, the most important mathematical discovery of the ancient science in the field of Harmony. This discovery at that time was called *"dividing a segment into the extreme and mean ratio"* [32].

From Refs. [44, 45], it is evident that prominent thinkers, scientists and mathematicians took part in the development of the *Mathematics of Harmony* for several millennia: Pythagoras, Plato, Euclid, Fibonacci, Pacioli, Kepler, Cassini, Binet, Lucas, Klein, and in the 20th century the well-known mathematicians Coxeter [7], Vorobyov [8], Hoggatt [9], Vaida [11], Knuth [123], and so on. And we cannot ignore this historical fact.

Fibonacci Numbers

The Fibonacci numbers, introduced into Western European mathematics in the 13th century by the Italian mathematician Leonardo of Pisa (known by the nickname Fibonacci), are closely related to the *golden ratio*. Fibonacci numbers from the numerical sequence, which starts with two units, and then each subsequent Fibonacci number is the sum of the two previous ones: $1, 1, 2, 3, 5, 8, 13, 21, 34, 55, \ldots$. The ratio of the two neighboring Fibonacci numbers in the limit tends to be the *golden ratio*.

The mathematical theory of Fibonacci numbers has been further developed in the works of the French mathematicians of the 19th century Binet (*Binet formula*) and Lucas (*Lucas numbers*). As mentioned above, in the second half of the 20th century, this theory was developed in the works of the Canadian geometer, Donald Coxeter [7], the Soviet mathematician, Nikolay Vorobyov [8], the American mathematician, Verner Hoggatt [9] and the English mathematician, Stefan Vajda [11], the outstanding American mathematician, Donald Knuth [123], and so on.

The development of this direction ultimately led to the emergence of the *Mathematics of Harmony* [6], a new interdisciplinary direction of modern science that relates to modern mathematics, computer science, economics, as well as to all theoretical natural sciences. The works of the well-known mathematicians, Coxeter [7], Vorobyov [8], Hoggatt [9], Vaida [11], Knuth [123], and others, as well as the study of Fibonacci mathematicians, members of the American Fibonacci Association, became the beginning of the process of *Harmonization of Mathematics* [68], which continues actively in the 21st century. And this process is confirmed by a huge number of books and articles in the field of the *golden section* and *Fibonacci numbers* published in the second half of the 20th century and the beginning of the 21st century [1–100].

Sources of the Present Three-Volume Book

The differentiation of modern science and its division into separate spheres do not allow us often to see the general picture of science and the main trends in its development. However, in science, there exist research objects that combine disparate scientific facts into a single whole. *Platonic solids* and the *golden section* are attributed to the category of such objects. The ancient Greeks elevated them to the level of *"the main harmonic figures of the Universe"*. For centuries or even millennia, starting from Pythagoras, Plato and Euclid, these geometric objects were the object of admiration and worship of the outstanding minds of mankind, during Renaissance, Leonardo da Vinci, Luca Pacoli, Johannes Kepler, in the 19th century, Zeising, Lucas, Binet and Klein. In the 20th century, the interest in these mathematical objects increased significantly, thanks to the research of the Canadian geometer, Harold Coxeter [7], the Soviet mathematician Nikolay Vorobyov [8] and the American mathematician Verner Hoggatt [9], whose works in the field of the Fibonacci numbers began the process of the "Harmonization of Mathematics". The development of this direction led to the creation of the *Mathematics of Harmony* [6] as a new interdisciplinary trend of modern science.

The newest discoveries in the various fields of modern science, based on the *Platonic solids*, the *golden section* and the *Fibonacci numbers*, and new scientific discoveries and mathematical results, related to the *Mathematics of Harmony* (*quasicrystals* [115], *fullerenes* [116], the new geometric theory of phyllotaxis (*Bodnar's geometry*) [28], the *general theory of the hyperbolic functions* [75, 82], the *algorithmic measurement theory* [16], the *Fibonacci and golden ratio codes* [6], the *"golden" number theory* [94], the *"golden" interpretation of the special theory of relativity* and the *evolution of the Universe* [87], and so on) create an overall picture of the movement of modern science towards the *"golden" scientific revolution*, which is one of the characteristic trends in the development of modern science. The sensational information about the experimental discovery of the golden section in the quantum world as a result of many years of research, carried out at the Helmholtz–Zentrum Berlin für Materialien und Energie (HZB) (Germany), the Oxford and Bristol Universities and the Rutherford Appleton Laboratory (UK), is yet another confirmation of the movement of the theoretical physics to the *golden section* and the *Mathematics of Harmony* [6].

For the first time, this direction was described in the book by Stakhov A.P., assisted by Scott Olsen, *The Mathematics of Harmony. From Euclid to Contemporary Mathematics and Computer Science*, World Scientific, 2009 [6].

In 2006, the Russian Publishing House, "Piter" (St. Petersburg) published the book, *Da Vinci Code and Fibonacci numbers* [46] (Alexey Stakhov, Anna Sluchenkova and Igor Shcherbakov were the authors of the book). This book was one of the first Russian books in this field. Some aspects of this direction are reflected in the following authors' books, published by Lambert Academic Publishing (Germany) and World Scientific (Singapore):

- Alexey Stakhov, Samuil Aranson, *The Mathematics of Harmony and Hilbert's Fourth Problem. The Way to Harmonic Hyperbolic and Spherical Worlds of Nature.* Germany: Lambert Academic Publishing, 2014 [51].
- Alexey Stakhov, Samuil Aranson, Assisted by Scott Olsen, *The "Golden" Non-Euclidean Geometry*, World Scientific, 2016 [52].
- Alexey Stakhov, *Numeral Systems with Irrational Bases for Mission-Critical Applications*, World Scientific, 2017 [53].

These books are fundamental in the sense that they are the first books in modern science, devoted to the description of the theoretical foundations and applications of the following new trends in modern science: the *history of the golden section* [78], the *Mathematics of Harmony* [6], the *"Golden" Non-Euclidean geometry* [52], ascending to Euclid's *Elements*, and also the *Numeral Systems with Irrational bases*, ascending to the Babylonian positional numeral system, the decimal and binary system and Bergman's system [54].

These books discuss the problems, which in modern mathematics are considered long resolved and therefore are not included in the circle of the studies of mathematicians, namely the new mathematical theory of measurement called the *Algorithmic Measurement Theory* [16, 17], the *Mathematics of Harmony* [6] as a new kind of elementary mathematics that has a direct relationship to the foundations of the mathematics and mathematical education, the new class of the elementary functions called the *hyperbolic Fibonacci and Lucas functions* [64, 75, 82] and finally, the new ways of real numbers representation, and the new binary and ternary arithmetic's [55, 72], which have the fundamental interest for computer science and digital metrology.

In 2010, the Odesa I.I. Mechnikov National University (Ukraine) hosted the *International Congress on the Mathematics of Harmony*. The main goal of the Congress was to consolidate the priority of Slavic science in the development of this important trend and acquaint the scientific community with the main trends of the development of the Mathematics of Harmony as the new interdisciplinary direction of modern science.

In the recent years, the new unique books on the problems of Harmony and the history of the golden section have been published:

• *The Prince of Wales. Harmony. A New Way of Looking at our World* (*coauthors Tony Juniper and Ian Skelly*). An Imprint of HarperCollins Publisher, 2010 [49].

• Hrant Arakelian, *Mathematics and History of the Golden Section*. Moscow, Publishing House "Logos", 2014 [50].

For the last 30 years, *Charles, The Prince of Wales*, had been known around the world as one of the most forceful advocates for the environment. During that period, he focused on many different aspects of our lives, when we continually confront with the real life from new angles of view and search original approaches. Finally, in *Harmony* (2010) [49], The Prince of Wales and his coauthors laid out their thoughts on the planet, by offering an in-depth look into its future. Here, we see a dramatic call to the action and an inspirational guide on the relationship of mankind with Nature throughout history. The Prince of Wales's *Harmony* (2010) [49] is an illuminating look on how we must reconnect with our past in order to take control of our future.

The 2014 book [50] by the Armenian philosopher and physicist *Hrant Arakelian* is devoted to the *golden section* and to the complexity of problems connected with it. The book consists of two parts. The first part is devoted to the mathematics of the golden section and the second part to the history of the golden section. Undoubtedly, Arakelian's 2014 book is one of the best modern books devoted to mathematics and the history of the golden section.

The *International Congress on Mathematics of Harmony* (Odessa, 2010) and the above-mentioned books by *The Prince of Wales* and Armenian philosopher *Hrant Arakelian* are brilliant confirmation of the fact that in modern science, the interest in the mathematics of the golden section and its history increases and further development of the Mathematics of Harmony can lead to revolutionary transformations in modern mathematics and science on the whole.

Why did the author decide to write the three-volume book *The Mathematics of Harmony as a New Interdisciplinary Direction and "Golden" Paradigm of Modern Science?* It should be noted that the author and other famous authors in this field published many original books and articles in this scientific direction. However, all the new results and ideas, described in the above-mentioned publications of Alexey Stakhov, Samuil Aranson, Charles, The Prince of Wales, Hrant Arakelian and other authors are scattered in their numerous articles and books, which makes it difficult to understand their fundamental role in the development of the modern mathematics, computer science and theoretical natural sciences on the whole.

This role is most clearly reflected in the following citations taken from *Harmony* by the Prince of Wales (2010) [49]:

> *"This is a call to revolution. The Earth is under threat. It cannot cope with all that we demand of it. It is losing its balance and we humans are causing this to happen."*

The following quote, placed on the back cover of Prince of Wales's *Harmony* [49], develops this thought:

> *"We stand at an historical moment; we face a future where there is a real prospect that if we fail the Earth, we fail humanity. To avoid such an outcome, which will comprehensively destroy our children's future or even our own, we must make choices now that carry monumental implications."*

Thus, *The Prince of Wales* has considered his 2010 book, *Harmony. A New Way of Looking at our World*, as a call to the revolution in modern science, culture and education. The same point of view is expressed in the above-mentioned books by Stakhov and Aranson [6, 46, 51–53]. Comparing the books of *Prince of Wales* [49] and *Hrant Arakelian* [50] to the 2009, 2016 and 2017 books of Alexey Stakhov and Samuil Aranson [6, 51–53], one can only be surprised how deeply all these books, written in different countries and continents, coincide in their ideas and goals.

Such an amazing coincidence can only be explained by the fact that in modern science, there is an urgent need to return to the "harmonious ideas" of Pythagoras, Plato and Euclid that permeated across the ancient Greek science and culture. The *Harmony* idea, formulated in the works of the Greek scholars and reflected in Euclid's *Elements* turned out to be immortal!

We can safely say that the above-mentioned books by Stakhov and Aranson (2009, 2016, 2017) [6, 51–53], the book by The Prince of Wales with the coauthors (2010) [49] and book by Arakelian (2014) [50] are the beginning of a revolution in modern science. The essence of this revolution consists, in turning to the fundamental ancient Greek idea of the *Universal Harmony*, which can save our Earth and humanity from the approaching threat of the destruction of all mankind.

It was this circumstance that led the author to the idea of writing the three-volume book *Mathematics of Harmony as a New Interdisciplinary Direction and "Golden" Paradigm of Modern Science*, in which the most significant and fundamental scientific results and ideas, formulated by the author and other authors (The Prince of Wales, Hrant Arakelian, Samuil Aranson and others) in the process of the development of this scientific direction, will be presented in a popular form, accessible to students of universities and colleges and teachers of mathematics, computer science, theoretical physics and other scientific disciplines.

Structure and the Main Goal of the Three-Volume Book

The book consists of three volumes:

- *Volume I. The Golden Section, Fibonacci Numbers, Pascal Triangle and Platonic Solids.*
- *Volume II. Algorithmic Measurement Theory, Fibonacci and Golden Arithmetic and Ternary Mirror-Symmetrical Arithmetic.*
- *Volume III. The "Golden" Paradigm of Modern Science: Prerequisite for the "Golden" Revolution in the Mathematics, the Computer Science, and Theoretical Natural Sciences.*

Because the *Mathematics of Harmony* goes back to the "harmonic ideas" of Pythagoras, Plato and Euclid, the publication of such a three-volume book will promote the introduction of these "harmonic ideas" into modern education, which is important for more in-depth understanding of the ancient conception of the *Universal Harmony* (as the main conception of ancient Greek science) and its effective applications in modern mathematics, science and education.

The main goal of the book is to draw the attention of the broad scientific community and pedagogical circles to the *Mathematics of Harmony*, which is a new kind of elementary mathematics and goes back to Euclid's *Elements*. The book is of interest for the modern mathematical education and can be considered as the "golden" paradigm of modern science on the whole.

The book is written in a popular form and is intended for a wide range of readers, including schoolchildren, school teachers, students of colleges and universities and their teachers, and also scientists of various specializations, who are interested in the history of mathematics, Platonic solids, golden section, Fibonacci numbers and their applications in modern science.

Introduction

What are "Paradigm" and "Scientific Revolution"?

As it is known, the term *paradigm* is derived from the Greek word *paradeigma* (example, sample) and refers to a combination of explicit and implicit (and often not realized) prerequisites that define the main essence of scientific research at some stage of scientific development.

This concept, in the modern sense of this term, was introduced by the American physicist and historian of science Thomas Kuhn (1922–1996) in the 1962 book *The Structure of Scientific Revolutions* [139]. According to Thomas Kuhn, a *paradigm* means a set of fundamental scientific ideas, which unite members of the scientific community and, conversely, the scientific community consists of people, who recognize the certain paradigm. As a rule, the *paradigm* is fixed in the textbooks and works of scientists and over the years determines the circle of problems and methods of their solution in a particular field of science. According to Kuhn [139], the examples of paradigms are Aristotle's views on education and ethics, Newtonian mechanics, etc.

A *paradigm shift* is a term also first introduced by Thomas Kuhn [139] for the description of changes in the basic assumptions within the framework of the leading theory in science (paradigm). Usually, a change of the scientific paradigm relates to the most dramatic events in the history of science. When a scientific discipline changes one paradigm for another, this is called the *scientific revolution* or *paradigm shift*, according to Kuhn's terminology [139]. The decision to abandon the *old paradigm* is always at the same time the decision to adopt the *new paradigm*; the proposal, which leads to such a decision, includes both a comparison of both paradigms with Nature and a comparison of the paradigms with each other.

What is the "Golden" Paradigm?

To answer this question, let us turn once again to the well-known statement by the genius of Russian philosophy, the aesthetics researcher of ancient Greece and the Renaissance, Alexei Losev (1893–1988), which is given in the Preface. In this statement, Losev in a very distinct form had formulated the essence of the *"golden"* *paradigm* of the ancient cosmology [5]:

> *"From Plato's point of view, and in general from the point of view of all the ancient cosmology, the world is some proportional Whole, obeying the law of harmonic division, the golden section (that is, the whole relates to the larger part, as the larger part to the smaller one)."*

In Losev's well-known statement [5], the essence of the *"golden"* *paradigm* of the ancient cosmology is formulated as follows. The *"golden"* *paradigm* is based on the most important ideas of ancient science, which in modern science are sometimes interpreted as *a curious result of unrestrained and wild fantasy* [5]. First of all, these are the *Pythagorean doctrine of the numerical harmony of the Universe* and *Plato's Cosmology*, based on the *golden section* and *Platonic solids*.

It is important to emphasize that Losev put the *golden section* in the center of the *"golden"* *paradigm* of ancient science. Thus, by referring to the geometric relations and geometric concepts, which expressed the *Universal Harmony*, in particular, the *golden section* and *Platonic solids*, Plato, along with Pythagoras, anticipated the emergence of mathematical natural sciences, which began developing rapidly in the 20th century. *The idea of Pythagoras and Plato about the Universe Harmony proved to be immortal.*

The Relationship Between Scientific Paradigms in Mathematics and Mathematical Natural Sciences

One of the original contemporary ideas, expressed in the article "Pseudoscience: a disease that there is no one to cure", written in 2011 by the talented Russian philosopher Denis Kleschev [128], is the fact that the *processes of paradigm shift in mathematics and natural sciences are closely interrelated.*

Kleschev notes as follows in [128]:

> *"Studying a history only for the sake of studying the history itself can hardly attract the attention of other researchers to it. Therefore, Kuhn's concept must be supplemented by consideration of both the internal and external structures of the change of scientific paradigms. To cope with this task is impossible if we are interested in the natural sciences in isolation from the study of the history of mathematics, as practiced by Thomas Kuhn. But if we include into the consideration the history of mathematics, rich with dramatic events and crises, as it immediately becomes apparent that to every paradigm leap in physics was preceded by cardinal changes in mathematics, preparing the ground for changing the natural science paradigm."*

The examples of the successful usage of *Platonic solids*, the *golden section*, and *Fibonacci numbers* in modern theoretical natural sciences, considered in Vols. I and II, allow expression of the idea that the process of harmonization of natural sciences has been realized actively in modern theoretical natural sciences. *Fullerenes* and *quasi-crystals*, awarded by the Nobel Prizes, are the most prominent examples of such harmonization, and this process requires a corresponding response from mathematics.

The development of modern *Fibonacci numbers theory* [7–9, 11] is a convincing example of *harmonization of mathematics*. This process obtained further reflection in Stakhov's book *The Mathematics of Harmony. From Euclid to Contemporary Mathematics and Computer*

Science [6] as a new interdisciplinary direction of modern science and mathematics.

Mathematization of Harmony and Harmonization of Mathematics

By considering the history of the development of mathematics since the ancient Greeks to the present time, we can distinguish the two processes that are closely related to each other, despite the more than 2000-year time distance between them. This connection is carried out through the *"golden"* *paradigm* of ancient Greeks as a fundamental conception that permeates the entire history of science. The first of these is the process of *Mathematization of Harmony.* This process began developing in ancient Greece in the sixth or fifth century BC (Pythagoras and Plato's mathematics) and ended in the third century BC by creating the greatest mathematical work of the ancient era, the Euclidean *Elements.* All efforts of the ancient Greeks were aimed at creating the mathematical doctrine of Nature, in the center of which the ancient Greeks placed the *Idea of Harmony,* which, according to *Proclus hypothesis,* had been expressed in the Euclidean *Elements* through *Platonic solids* (XIIIth Book of the Elements) and the *golden section* (Book II, Proposition of II.11).

The process of *Mathematization of Harmony* in the ancient period ended with the creation of Euclidean *Elements;* the main purpose of this process was the creation of the complete geometric theory of *Platonic solids* (Book XIII of the *Elements*), which expressed the *Universal Harmony* in *Plato's cosmology.* To create this theory, Euclid already in Book II introduced the *task of dividing a segment in extreme and mean ratio* (the Euclidian name for the *golden section*), which was used by Euclid by creating the geometric theory of the *dodecahedron,* based on the *golden ratio.*

Harmonization of Mathematics is a process opposite to *Mathematization of Harmony* [68]. This process began developing most rapidly in the second half of the 20th century in the works of the Canadian geometer Harold Coxeter [7], the Soviet mathematician Nikolay Vorobyov [8], the American mathematician

Verner Hoggatt [9], the English mathematician Stefan Waida [11], and other famous Fibonacci mathematicians. The creators of the modern *Fibonacci number theory* [7–9, 11] have acted very wisely and cautiously, not attracting attention to the fact that *Fibonacci numbers* are one of the most important numerical sequences, which together with the *golden section* actually express *Harmony of Nature*. They "euthanized" the vigilance of modern orthodox mathematicians, which allowed them to establish the *Fibonacci Association*, the mathematical journal *The Fibonacci Quarterly* and, starting from 1984, regularly (once every 2 years) holding the International Conference on *Fibonacci Numbers and their Applications*. Thanks to the active work of the *Fibonacci Association*, it was possible to combine the efforts of a huge number of researchers, who found the *Fibonacci numbers* and the *golden ratio* in their scientific areas. Starting from the last decade of the 20th century, the so-called *Slavic Golden Group* began playing an active role in the development of this direction. The *Slavic Golden Group* was established in Kiev (the capital of Ukraine) in 1992 during the First International Workshop *Golden Proportion and Problems of Harmony Systems*. This scientific group included leading scientists and lovers of the *golden ratio* and *Fibonacci numbers* from Ukraine, Russia, Belarus, Poland, Armenia and other countries.

In 2003, according to the initiative of the *Slavic Golden Group*, the International Conference on *Problems of Harmony, Symmetry and the Golden Section in Nature, Science and Art* was held at Vinnitsa Agrarian University by the initiative of Professor Alexey Stakhov. According to the decision of the conference, the *Slavic Golden Group* was transformed into the *International Club of the Golden Section*.

In 2005, the *Golden Section Institute* was organized at the Academy of Trinitarism (Russia). In 2010, according to the initiative of the *International Club of the Golden Section*, the *First International Congress on Mathematics of Harmony* was held on the basis of the Odessa Mechnikov National University (Ukraine). All these provide evidence of the fact that the *International Club of the Golden Section* plays in the Russian-speaking scientific community the same role as the American Fibonacci Association in the English-speaking scientific community. The publication of Stakhov's book *The Mathematics of Harmony. From Euclid to Contemporary Mathematics and Computer Science* (World Scientific, 2009) [6] was an *important event in harmonization of modern science and mathematics.*

What is Harmonization of Mathematics?

This, first of all, refers to the wide use of fundamental concept of *Mathematics of Harmony*, such as the *Platonic Solids*, the *golden proportion*, the *Fibonacci numbers* and their generalizations (the *Fibonacci p-numbers*, the *metallic proportions or the "golden" p-proportions*, etc.), as well as new mathematical concepts (the *Fibonacci matrices*, the *"golden" matrices*, the *hyperbolic Fibonacci and Lucas functions* [64, 75], etc.) to solve certain mathematical problems and create new mathematical theories and models.

The brilliant examples are the solution of *Hilbert's 10th Problem* (Yuri Matiyasevich, 1970), based on the use of new mathematical properties of the *Fibonacci numbers*, and the solution of *Hilbert's Fourth problem* (Alexey Stakhov and Samuil Aranson), based on the use of Spinadel's *metallic proportions*. The theory of *numeral systems with irrational bases* (*Bergman's system* and the *codes of the golden ratio*) and the concept of the *"golden" number theory*, arising from them, are examples of the original and far from trivial mathematical results, obtained in the framework of *Mathematics of Harmony* [6].

The main merit of the modern mathematicians in the field of *golden ratio and Fibonacci numbers* consisted in the fact that their researches *caused the spark, from which the flame had ignited.* The process of *Harmonization of Mathematics* is confirmed by a rather impressive and far from complete list of modern books in this field, published in the second half of the 20th century and early 21st century [1–53].

Among them, the following three books, published in the 21st century, deserve special attention:

(1) **Stakhov Alexey. Assisted by Scott Olsen.** *The Mathematics of Harmony. From Euclid to Contemporary Mathematics and Computer Science* (World Scientific, 2009) [6]

(2) **The Prince of Wales with co-authors.** *Harmony. A New Way of Looking at our World* (New York: Harpert Collins Publishers, 2010) [51]

(3) **Arakelyan Hrant.** *Mathematics and History of the Golden Section* (Moscow: Logos, 2014) [50] (Russian).

What Place Does Mathematics of Harmony Occupy in the System of Modern Mathematical Theories?

To answer this question, it is appropriate to consider a quote from the review of the prominent Ukrainian mathematician, academician Yuri Mitropolskiy on the scientific research of the Ukrainian scientist Professor Alexey Stakhov. In this review, Yuri Mitropolskiy reports the following:

> *"I have followed the scientific career of Professor Stakhov for a long time — seemingly since the publication of his first 1977 book, "Introduction into Algorithmic Measurement Theory", which was presented by Professor Stakhov in 1979 at the scientific seminar of the Mathematics Institute of the Ukrainian Academy of Sciences. I became especially interested in Stakhov's scientific research after listening his brilliant speech at the 1989 session of the Presidium of the Ukrainian Academy of Sciences. In his speech, Professor Stakhov reported on scientific and engineering results in the field of 'Fibonacci computers' that were conducted under his scientific supervision at the Vinnitsa Technical University...*
>
> *One may wonder what place does take this work in the general theory of mathematics. As it seems to me, that in the last few centuries, as Nikolay Lobachevsky said, "Mathematicians have turned all their attention to the Advanced parts of analytics, and by neglecting the origins of mathematics and did not wishing to work in that field, which they passed and left behind. As a result, it was created a gap between 'Elementary Mathematics', the basis of modern mathematical education, and 'Advanced Mathematics.' In my opinion, the Mathematics of Harmony, developed by Professor Stakhov, fills out that gap. The Mathematics of Harmony is a big theoretical contribution to the development of the 'Elementary Mathematics', and the Mathematics of Harmony should be considered as great contribution to mathematical education."*

Note that Alexey Stakhov used Mitropolsky's review as the Preface to Stakhov's 2009 book *The Mathematics of Harmony* [6].

Thus, academician Mitropolsky in his review focuses on the historical aspect. His point of view is that Stakhov's *Mathematics of Harmony* is, first of all, a new kind of *elementary mathematics*, based on the unusual interpretation of the Euclidean *Elements*, as historically the first version of *Mathematics of Harmony*, connected with the *Platonic solids* and the *golden section*.

But besides this, there are other aspects of *Mathematics of Harmony*: *applied* and *aesthetical*. First of all, we should note the *applied nature* of *Mathematics of Harmony*, which is the true *Mathematics of Nature*. *Mathematics of Harmony* is found in many natural phenomena, such as the *movement of Venus across the sky* (*"Pentacle of Venus"*), the *pentagonal symmetry in Nature*, the *botanical phenomenon of phyllotaxis*, the *fullerenes*, the *quasicrystals*, etc.

On the other hand, *Mathematics of Harmony* [6] by the name and by the contents fully satisfies *Hutcheson and Dirac principles of mathematics beauty* [154]. According to Dirac, *the main mathematical ideas should be expressed in terms of excellent mathematics*. This means that *Mathematics of Harmony*, which was aroused in the ancient Greek mathematics, is a *beautiful mathematics*, which must be embodied in the structures of Nature and contemporary science. This conclusion is confirmed by the modern scientific achievements, described in Vols. I and II and will be discussed in detail in this volume.

The harmonious combination of the *applied aspect* of *Mathematics of Harmony*, as the true *Mathematics of Nature*, with its aesthetic perfection (*Hutcheson and Dirac principles* [154]), gives us reason to suggest that it is *Mathematics of Harmony* that can become the *"golden" paradigm of modern science*, which will help overcome the crisis in modern mathematics [101]. *Mathematics of Harmony*, described in this three-volume book, is a very young mathematical theory, although in its origins it goes back to the Euclidean *Elements*.

The term *Mathematics of Harmony* was used first by Alexey Stakhov in the speech *The Golden Section and Modern Harmony Mathematics* [66], made in 1996 at the Seventh International Conference on *Fibonacci Numbers and Its Applications* (Austria, Graz, 1996).

The speech was perceived with great interest by the Fibonacci mathematicians, as evidenced by the fact that this speech was selected for publication in the collection of papers of the International Conference on *Applications of Fibonacci Numbers*, published by Kluwer Academic Publishers in 1998 [66]. Starting from this publication, the development of *Mathematics of Harmony* became the focus of Alexey Stakhov's scientific interests, which led him to the publication of the book [6] and to the writing of this three-volume book.

The Goal of Vol. III

The main goal of Vol. III is to answer the following two questions:

(1) What place does Mathematics of Harmony occupy in the system of contemporary mathematical sciences and how does it influence the development of modern science and mathematics?
(2) Is *Mathematics of Harmony* the *"golden"* *paradigm* of modern science?

Volume III begins with discussion on the influence of *Mathematics of Harmony* on the course of the development of modern mathematics and computer science; a number of unusual ideas put forward in the first two volumes of this book. In particular, *Proclus hypothesis*, which was discussed in Volume I, is considered as a prerequisite to the *"golden"* *revolution* in the history of mathematics.

The influence of *Mathematics of Harmony* [6] on the development of two of the most ancient mathematical theories, the *measurement theory* and the *elementary theory of numbers*, is discussed. Next, the *numeral systems with irrational bases* (the *Fibonacci codes* and the *codes of the golden proportion*) are considered as a prerequisite for the "golden" revolution in computer science, as well as the elements of the "golden" theory of numbers, based on the *golden ratio*. In conclusion, the article by the famous Russian philosopher Sergey Abachiev *"Mathematics of Harmony through the Eyes of the Historian and Expert of Methodology of Science"* [156] is discussed.

Chapter 2 introduces a new class of hyperbolic functions, based on the classical *golden proportion* and its generalization, the *golden p-proportions*.

Chapter 3 is devoted to the discussion of the connection between *Mathematics of Harmony* and the *Theory of elementary functions*, which plays a fundamental role in mathematics and its applications in theoretical natural sciences. Here, a new class of "elementary functions" is introduced: the *"golden" hyperbolic functions* or the *hyperbolic Fibonacci and Lucas functions* [57, 58].

Chapter 4 discusses the applications of the *"golden" hyperbolic functions* in the *new geometric theory of phyllotaxis*, created by the Ukrainian researcher Oleg Bodnar [28], and also the function *Golden Shofar* and *Shofar-like model of the Universe* [77].

Chapter 5 is devoted to outlining the theory of *Fibonacci numbers*, which is the result of the collective creativity of several researchers from different countries and continents: Vera de Spinadel, Argentina [29]; Midhat Gazale, France [30]; Alexander Tatarenko, Russia [62]; Jay Kappraff, USA [33, 34]; Grant Arakelyan, Armenia [49, 63]; Victor Shenyagin, Russia [64]; Nikolay Kosinov, Ukraine [65]; Alexey Stakhov, Canada [66]; Spears, Bicknell-Johnson [67]), and others.

Chapter 6 is the central chapter from the point of view of the answer to the questions posed at the beginning of this Introduction. Chapter 6 addresses a wide range of issues relating to mathematics and its history. The crisis in modern mathematics, described in the book of the outstanding American historian of mathematics Morris Klein *Mathematics. The Loss of Certainty* [51], is analyzed. Further, in Chapter 6, special attention is paid to the analysis of "strategic mistakes" in the development of mathematics, described in Stakhov's articles [71, 72]. The criteria of aesthetics and beauty of mathematics are considered, in particular, the *Dirac principle of mathematical beauty*. From the standpoint of these criteria, the most important mathematical results, obtained in the framework of *Mathematics of Harmony* [6, 46, 47], are analyzed.

In Chapter 6, special attention is paid to the analysis of the strategic errors in the development of mathematics conducted in Stakhov's article [71]. The criteria of aesthetics and beauty of mathematics, in particular, *Dirac's principle of mathematical beauty* are considered. From the standpoint of these criteria, the most important mathematical results, obtained in the framework of Stakhov's 2009 book *Mathematics of Harmony* [6], are analyzed.

Mathematics of Harmony is discussed as the "golden" paradigm of modern science and also the interrelation of changes of the scientific paradigms in mathematics and theoretical natural sciences, and an attempt is made to answer the question about the place of *Mathematics of Harmony* in the system of the modern mathematical sciences.

About the Author

Alexey Stakhov, born in May 7, 1939, is a Ukrainian mathematician, inventor and engineer, who has made a contribution to the theory of Fibonacci numbers and the *golden section* and their applications in computer science and measurement theory. He is a Doctor of Computer Science (1972) and a Professor (1974), and the author of over 500 publications, 14 books and 65 international patents. He is also the author of many original publications in computer science and mathematics, including *algorithmic measurement theory* [16, 17], *Fibonacci codes and codes of the golden proportions* [19], *hyperbolic Fibonacci and Lucas functions* [64, 75] and finally the *Mathematics of Harmony* [6], which goes back in its origins to Euclid's *Elements*. In these areas, Alexey Stakhov has written many papers and books, which have been published in famous scientific journals by prestigious international publishers.

The making of Alexey Stakhov as a scientist is inextricably linked with the Kharkov Institute for Radio Electronics, where he was a postgraduate student of the Technical Cybernetics Department from 1963 to 1966. Here, he defended his PhD thesis in the field of Technical Cybernetics (1966) under the leadership of the prominent Ukrainian scientist Professor Alexander Volkov. In 1972, Stakhov defended (at the age of 32 years) his Grand Doctoral dissertation *Synthesis of Optimal Algorithms for Analog-to-Digital Conversion* (Computer Science speciality). Although the dissertation had an engineering character, Stakhov in his books and articles touched upon two fundamental problems of mathematics: *theory of measurement* and *numeral systems*.

Prof. Stakhov worked as "Visiting Professor" of different Universities: Vienna Technical University (Austria, 1976), University of Jena (Germany, 1986), Dresden Technical University (Germany, 1988), Al Fateh University (Tripoli, Libya, 1995–1997), Eduardo Mondlane University (Maputo, Mozambique, 1998–2000).

Stakhov's Prizes and Awards

- Award for the best scientific publication by Ministry of Education and Science of Ukraine (1980);
- Barkhausen's Commemorative Medal issued by the Dresden Technical University as "Visiting Professor" of Heinrich Barkhausen's Department (1988);
- Emeritus Professor of Taganrog University of Radio Engineering (2004);
- The honorary title of "Knight of Arts and Sciences" (Russian Academy of Natural Sciences, 2009);
- The honorary title "Doctor of the Sacred Geometry in Mathematics" (American Society of the Golden Section, 2010);
- Awarded "Leonardo Fibonacci Commemorative Medal" (Interdisciplinary Journal "De Lapide Philosophorum", 2015).

Acknowledgments

Alexey Stakhov expresses great thanks to his teacher, the outstanding Ukrainian scientist, Professor Alexander Volkov; under his scientific leadership, the author defended PhD dissertation (1966) and then DSc dissertation (1972). These dissertations were the first steps in Stakhov's research, which led him to the conceptions of *Mathematics of Harmony* and *Fibonacci computers* based on the *golden section* and *Fibonacci numbers*.

During his stormy scientific life, Stakhov met many fine people, who could understand and evaluate his enthusiasm and appreciate his scientific direction. About 50 years ago, Alexey Stakhov had read the remarkable brochure *Fibonacci Numbers* [8] written by the famous Soviet mathematician Nikolay Vorobyov. This brochure was the first mathematical work on, Fibonacci numbers published in the second half of the 20th century. This brochure, determined Stakhov's scientific interest in the *Fibonacci numbers* and the *golden section* for the rest of his life. In 1974, Professor Stakhov met with Professor Vorobyov in Leningrad (now St. Petersburg) and told Professor Vorobyov about his scientific achievements in this area. Professor Vorobyov gave Professor Stakhov, his brochure *Fibonacci Numbers* [8] as a keepsake with the following inscription: "*To highly respected Alexey Stakhov with Fibonacci's greetings*". This brief inscription because a certain kind of guiding star for Alexey Stakhov.

With deep gratitude, Stakhov recollects the meeting with the famous Austrian mathematician Professor *Alexander Aigner* in the Austrian city of Graz in 1976. The meeting with Professor Aigner was the beginning of the international recognition of Stakhov's scientific direction.

Another remarkable scientist, who had a great influence on Stakhov's research, was the Ukrainian mathematician and academician *Yuri Mitropolskiy*, the *Head of the Ukrainian Mathematical School* and the *Chief Editor of the Ukrainian Mathematical Journal*. His influence on Stakhov's researches, pertinent to the history of mathematics and other topics, such as the application of the Mathematics of Harmony in contemporary mathematics, computer science and mathematical education, were inestimable stimulus for Alexey Stakhov. Thanks to the support of Yuri Mitropolskiy, Stakhov published many important articles in the prestigious Ukrainian academic journals, including the *Ukrainian Mathematical Journal*.

In 2002, *The Computer Journal* (British Computer Society) published the fundamental article by Stakhov, *"Brousentsov's Ternary Principle, Bergman's Number System and Ternary Mirror-Symmetrical Arithmetic"* (*The Computer Journal*, Vol. 45, No. 2, 2002) [72]. This article by Stakhov created great interest among all the English scientific computer community. Emeritus Professor of Stanford University *Donald Knuth* was the first outstanding world scientist, who congratulated Prof. Stakhov with this publication. Donald Knuth's letter became one of the main stimulating factors for writing Stakhov's future book *The Mathematics of Harmony From Euclid to Contemporary Mathematics and Computer Science* (World Scientific, 2009) [6]. *Professor Stakhov considers this book as the main book of his scientific life.*

Stakhov's arrival to Canada in 2004 became the beginning of a new stage in his scientific research. Within 15 years (2004–2019), Prof. Stakhov had published more than 50 fundamental articles in different international English-language journals, such as *Chaos, Solitons & Fractals, Applied Mathematics, Arc Combinatoria, The Computer Journal, British Journal of Mathematics and*

Computer Science, Physical Science International Journal, Visual Mathematics, etc. Thanks to the support of Prof. El-Nashie, the Editor-in-Chief of the Journal *Chaos, Solitons & Fractals* (UK), Stakhov published in this journal 15 fundamental scientific articles that garnered great interest among the English-speaking scientific community.

The publication of the three fundamental books *The Mathematics of Harmony* (World Scientific, 2009) [6], *The "Golden" Non-Euclidean Geometry* (World Scientific, 2016, co-author Prof. Samuil Aanson) [52] and *Numeral Systems with Irrational Bases for Mission-Critical Applications* (World Scientific, 2017) [53] is one of the main scientific achievements by Stakhov during the Canadian period of his scientific creativity. These books were published thanks to the support of the famous American mathematician Prof. *Louis Kauffman*, Editor-in-Chief of the *Series on Knots and Everything* (World Scientific) and Prof. *M.S. Wong*, the famous Canadian mathematician (York University) and Editor-in-Chief of the *Series on Analysis, Application and Computation* (World Scientific). A huge assistance in the publication of Stakhov's books by of World Scientific was rendered by the American researcher, Scott Olsen, Professor of Philosophy at the College of Central Florida, and Jay Kappraff, Emeritus Professor of Mathematics at the New Jersey Institute of Technology. Prof. Scott Olsen, who was one of the leading US experts in the field of *Harmony* and the *golden section*, was the English editor for Stakhov's book mentioned above and the Emeritus Professor Jay Kappraff was the reviewer of Stakhov's book, *The Mathematics of Harmony* (World Scientific, 2009).

The prominent Ukrainian mathematician and head of the Ukrainian Mathematical School, *Yuri Mitropolskiy*, praised highly Stakhov's *Mathematics of Harmony*. Academician Mitropolsky organized Stakhov's speech at the meeting of the Ukrainian Mathematical Society in 1998. Based upon his recommendation, Stakhov's articles were published in the Ukrainian academic journals, in particular, the *Ukrainian Mathematical Journal*. Under his direct influence, Stakhov started writing the book, *The Mathematics of Harmony. From Euclid to Contemporary Mathematics and Computer*

Science [6], which was published by World Scientific in 2009 following the death of the academician Mitropolsky in 2008.

Scientific cooperation of Alexey Stakhov and Samuil Aranson

Samuil Aranson's acquaintance to the *golden section* and the *Fibonacci numbers* began in 2001 after the reading of a very rare book "Chain Fractions" [107] by the famous Russian mathematician, Aleksandr Khinchin. In this book, Samuil Aranson found results, related to the representation of the *"golden ratio"* in the form of a continued fraction.

In 2007, Prof. Aranson read a wonderful Internet publication, *Museum of Harmony and Golden Section*, posted in 2001 by Professor Alexey Stakhov and his daughter Anna Sluchenkova. This Internet Museum covers various areas of modern natural sciences and tells about the different and latest scientific discoveries, based on the *golden ratio* and *Fibonacci numbers*, including the *Mathematics of Harmony* and its applications in modern natural sciences. After reading this Internet Museum, Samuil Aranson contacted Alexey Stakhov in 2007 through e-mail and offered him joint scientific collaboration in further application of the *Mathematics of Harmony* in various areas of mathematics and modern natural sciences. Scientific collaboration between Alexey Stakhov and Samuil Aranson turned out to be very fruitful and continues up to the present time.

New ideas in the field of elementary mathematics and the history of mathematics, developed by Stakhov (*Proclus's hypothesis* as a new look at Euclid's *Elements* and history of mathematics, *hyperbolic Fibonacci and Lucas functions* [64, 75] as a new class of elementary functions and other mathematical results) attracted the special attention of Prof. Aranson. Scientific collaboration between Stakhov and Aranson began in 2007. From 2007, they published the following joint scientific works (in Russian and English), giving fundamental importance for the development of mathematics and modern theoretical natural sciences:

Stakhov and Aranson's Mathematical Monographs in English

1. Stakhov A., Aranson S., *The Mathematics of Harmony and Hilbert's Fourth Problem. The Way to the Harmonic Hyperbolic and Spherical Worlds of Nature.* Germany: Lambert Academic Publishing, 2014.
2. Stakhov A., Aranson S., Assisted by Scott Olsen, *The "Golden" Non-Euclidean Geometry: Hilbert's Fourth Problem, "Golden" Dynamical Systems, and the Fine-Structure Constant,* World Scientific, 2016.

Stakhov and Aranson's Scientific Papers in English

3. Stakhov A.P., Aranson S.Kh., "Golden" Fibonacci goniometry, Fibonacci-Lorentz transformations, and Hilbert's fourth problem. *Congressus Numerantium* **193**, (2008).
4. Stakhov A.P., Aranson S.Kh., Hyperbolic Fibonacci and Lucas functions, "golden" Fibonacci goniometry, Bodnar's geometry, and Hilbert's fourth problem. Part I. Hyperbolic Fibonacci and Lucas functions and "Golden" Fibonacci goniometry. *Applied Mathematics* **2**(1), (2011).
5. Stakhov A.P., Aranson S.Kh., Hyperbolic Fibonacci and Lucas functions, "golden" Fibonacci goniometry, Bodnar's geometry, and Hilbert's fourth problem. Part II. A new geometric theory of phyllotaxis (Bodnar's Geometry). *Applied Mathematics* **2**(2), (2011).
6. Stakhov A.P., Aranson S.Kh., Hyperbolic Fibonacci and Lucas functions, "golden" Fibonacci goniometry, Bodnar's geometry, and Hilbert's fourth problem. Part III. An original solution of Hilbert's fourth problem. *Applied Mathematics* **2**(3), (2011).
7. Stakhov A.P., Aranson S.Kh., The mathematics of harmony, Hilbert's fourth problem and Lobachevski's new geometries for physical world. *Journal of Applied Mathematics and Physics* **2**(7), (2014).

8. Stakhov A., Aranson S., The fine-structure constant as the physical-mathematical millennium problem. *Physical Science International Journal* **9**(1), (2016).

9. Stakhov A., Aranson S., Hilbert's fourth problem as a possible candidate on the millennium problem in geometry. *British Journal of Mathematics & Computer Science* **12**(4), (2016).

Chapter 1

Mathematics of Harmony as a Prerequisite for the "Golden" Revolution in Mathematics and Computer Science

1.1. "Proclus Hypothesis" as a Prerequisite for the "Golden" Revolution in the History of Mathematics

1.1.1. The significance of the "Proclus hypothesis" for the development of mathematics

This amazing hypothesis was discussed in detail in Vol. I. Its significance for the history of mathematics and its future development are difficult to overestimate. The main conclusion from the *Proclus hypothesis* reduces to the fact that the famous Euclidean *Elements*, the greatest ancient Greek mathematical work, where modern mathematics goes back in its origins, was written by Euclid under the direct influence of the ancient *Greek Idea of Harmony*, which was associated in *Plato's cosmology* with the *Platonic solids* (Alexei Losev).

According to this hypothesis, the *Pythagorean doctrine of the numerical harmony of the Universe* and *Plato's Cosmology*, based on regular polyhedra, was embodied in the Euclidean *Elements*, the most famous mathematical work of ancient Greek mathematics. From this point of view, we have every right to consider the Euclidean *Elements* as the first attempt to create the *Mathematical Theory of Universe Harmony. This fact is the main secret of Euclidean*

Elements and the most unusual view on Euclidean Elements, which led us to the revision of mathematics history, starting from ancient Greek mathematics.

Proclus hypothesis had a great influence on the development of science and mathematics. As stated in the books of the Western historians of mathematics [117, 121, 122], Johannes Kepler was convinced in the correctness of the *Proclus hypothesis*, and this is confirmed by his "Cosmic Cup", the original model of the solar system, based on the *Platonic solids.*

In the 19th century, the prominent German mathematician Felix Klein (1849–1925) suggested the hypothesis that the *icosahedron,* one of the most important *Platonic Solids,* is the main geometric object of mathematics, which allows uniting all the most important branches of mathematics: *geometry, Galois theory, group theory, invariant theory,* and *differential equations* [113]. Unfortunately, this idea of Klein did not receive further development in mathematics, which can also be considered as a *'strategic mistake'* in the history of mathematics.

1.1.2. Proclus hypothesis and the "golden" revolution in the history of mathematics

In Vol. I, we discussed a new look at the history of mathematics, which follows from the *Proclus hypothesis.* As it is known, the outstanding Russian mathematician academician Andrey Kolmogorov in his book [102] singled out the two main, that is, "key" problems of mathematics, which stimulated the development of ancient mathematics at the initial stage of its development: *counting problem* and *measurement problem.*

The essence of this problem consists in the fact that another additional "key" problem of mathematics arose from the *Proclus hypothesis*: the *Harmony problem,* which had a tremendous impact on the development of mathematics in subsequent periods of its development.

This problem was associated with the *Platonic solids* and the *golden section,* the most important mathematical discoveries of ancient mathematics, which expressed the *Universal harmony*. As it is known, the *Harmony Problem* was the central idea of Euclid's *Elements*; the main purpose of the *Elements*, according to the *Proclus hypothesis*, was to create the complete geometric theory of the *Platonic solids*, which in *Plato's cosmology* expressed the *Universe Harmony*.

This idea leads to a new look at the history of mathematics starting from Euclid's *Elements*. This look is detailed in Vol. I and further reflected here in Fig. 1.1.

The approach, following from the *Proclus hypothesis* and reflected in Fig. 1.1, leads to a conclusion, which can be a surprise for many mathematicians. It turns out that in parallel with *Classical Mathematics* in ancient mathematics, starting from ancient Greeks, another mathematical direction was developing: the *Mathematics of Harmony* [6]. This new mathematical direction, similar to *Classical Mathematics*, goes back in its origin to Euclid's *Elements*, but focuses not upon the *axiomatic approach* (Euclidean axioms), the *theory of irrationality*, and the *elementary numbers theory*, but upon the geometric *task of dividing a segment in extreme and mean ratio* (Euclidean name of the *golden section*, Proposition II.11) and on the theory of *Platonic solids*, described in Book XIII of the Euclidean

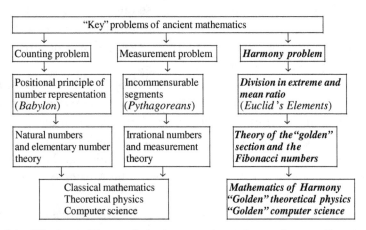

Fig. 1.1. "Key" problems of ancient mathematics and new directions in mathematics, theoretical physics, and computer science (taken from Vol. I).

Elements. Many outstanding thinkers and scientists took part in the creation of *Mathematics of Harmony* for more than 2000 years: Pythagoras, Plato, Euclid, Fibonacci, Pacioli, Kepler, Cassini, Binet, Lucas, Klein, and, in the 20th century, Vorobyov, Hoggatt, Waida and many others.

It is clear that if this truth triumphs, that is, the Proclus hypothesis is recognized by the world mathematical community as the MILLENIUM PROBLEM in mathematics history [95], then this historical truth will be the beginning of a new stage in the history of modern mathematics and mathematical education (this idea belongs to academician Yuri Mitropolskiy). Note that this idea is actually also described in many historical books of the Western historians of mathematics [117, 121, 122].

It is clear that the *Proclus hypothesis* can be considered a prerequisite for introducing the ancient concept of Harmony into modern mathematics. At the same time, the "idea of Harmony" introduces into contemporary mathematics the main applied goal, for the sake of which the ancient Greeks created *Mathematical Theory of Nature* [101], namely, the creation of new mathematical theories of theoretical natural sciences and of useful mathematical models of *Universal Harmony*; this means that the *Platonic solids*, the *golden section*, the *Fibonacci numbers* and their generalizations can be the basis of these theories and models. These outstanding mathematical discoveries of ancient and medieval mathematics will cease to play the role of some "outcasts" in mathematics and will occupy in it a worthy place that they certainly deserve.

This process, which is called in Ref. [68] as the *Harmonization of Mathematics*, can become the new *"golden" paradigm of modern science* and the beginning of the *"golden" revolution* in mathematics, which will bring about the merger of mathematics with theoretical natural sciences. Note that this is the main idea of Morris Klein, which he expounded in the final chapter, "The Authority of Nature", of his remarkable book [101].

The establishment of the American *Fibonacci Association* (1963), the mathematical journal *The Fibonacci Quarterly* (1963), the holding of the International Conference *Fibonacci Numbers and*

Their Applications (once every 2 years starting from 1984), the creation of the so-called *Slavic Golden Group* (Kiev, 1992), which in 2003 was transformed into the *International Club of the Golden Section*, the establishment of the *Institute of the Golden Section* (Russia, Academy of Trinitarism) (2005), the *International Congress on the Mathematics of Harmony* (Odessa, 2010), and finally, the publication of the first Stakhov book *The Mathematics of Harmony* [6] are all very important scientific events, which indicate that in modern mathematics, the process of the harmonization of modern mathematics and theoretical natural sciences, that is, the process of the revival of the "harmonious" ideas of Pythagoras, Plato and Euclid, goes on in full swing.

1.2. The Paradigm Shift to the "Golden" Elementary Number Theory

1.2.1. What is "elementary number theory"?

In the article "Number Theory" of the Russian language Wikipedia, we find the following answer to this question:

*"In the **elementary number theory**, the integers are studied without using the methods of other branches of mathematics. Among the main thematic areas of the elementary number theory we can select the following:*

- *The theory of the divisibility of integers.*
- *Euclidean algorithm for calculating the greatest common factor and the least common multiple.*
- *Decomposition of a number into prime factors and the main theorem of arithmetic.*
- *Theory of comparisons by modulo.*
- *Continued fractions.*
- *Diophantine equations, that is, the solution of indefinite equations in integers.*
- *Study of some classes of integers: perfect numbers, Fibonacci numbers, etc.*
- *Fermat's small theorem and its generalization: Euler's theorem.*
- *Finding Pythagorean triples, the problem of the four cubes.*

- *Entertaining mathematics, for example, the construction of magic squares."*

It is important *to emphasize that the Fibonacci numbers* are also included into the *elementary number theory.*

The "golden" number theory as a new paradigm shift to the elementary number theory: The new positional numeral systems and new computer arithmetics [19, 54, 55, 72] are the most important applications of *Mathematics of Harmony* for computer science. Let's consider these important scientific results, which signifies the **new paradigm shift** to the *elementary number theory,* detailed in the following sections.

1.2.2. Bergman's system

In 1957, the young American mathematician George Bergman published the article "A number system with an irrational base" in the authoritative journal *Mathematics Magazine* [54]. The following sum is called *Bergman's system*:

$$A = \sum_i a_i \Phi^i, \tag{1.1}$$

where A is any real number, a_i is a binary numeral $\{0,1\}$ of the ith digit, $i = 0, \pm 1, \pm 2, \pm 3, \dots$, Φ^i is the weight of the ith digit, and $\Phi = (1 + \sqrt{5})/2$ is the base of the numeral system (1.1).

On the face of it, there isn't an essential distinction between the formula (1.1) for *Bergman's system* and the formulas for the canonical positional numeral systems, in particular, the *binary system*:

$$A = \sum_i a_i 2^i (i = 0, \pm 1, \pm 2, \pm 3, \dots) \ (a_i \in \{0,1\}), \tag{1.2}$$

where the digit weights are connected by the following "arithmetical" relations:

$$2^i = 2^{i-1} + 2^{i-1} = 2 \times 2^{i-1}, \tag{1.3}$$

which underlie the *binary arithmetic.*

The principal distinction between *Bergman's system* (1.1) and the *binary system* (1.2) is the fact that the irrational number $\Phi = (1 + \sqrt{5})/2$ (the golden ratio) is used as the base of the numeral system (1.1) and its digit weights are connected by the following relationships:

$$\Phi^i = \Phi^{i-1} + \Phi^{i-2} = \Phi \times \Phi^{i-1}, \qquad (1.4)$$

which underlie the *"golden" arithmetic.*

This is why Bergman called this mathematical result (1.1) *the numeral system with irrational base.* Although Bergman's article [54] is a fundamental result for the elementary theory of numbers and computer science, the mathematical discovery of the genius American wunderkind proved to be so unexpected that many mathematicians and computer science experts of that period were not able to estimate the fundamental significance of this mathematical discovery. However, from Kuhn's point of view [139], we can make the following bold assumption: *Bergman's system is the "golden" paradigm shift to the elementary theory of numbers and computer science, which can have profound implications for further development of mathematics and computer science.*

1.2.3. New classes of the numeral systems with irrational bases

Further development of the concept of numeral systems with irrational base was given in the works of the author [19, 53, 99, 100] (see Table 1.1).

The new classes of numeral systems with irrational bases, presented in Table 1.1, can be interpreted as the *"golden" paradigm shift* toward the *elementary theory of numbers* and *computer science.* This *"golden" paradigm shift* creates new prospects in the development of computer technology and in designing the new noise-resistant and fault-tolerant computers and processors for mission-critical applications [99, 100].

As shown in Chapters 1 and 2 of Volume II, the so-called *Algorithmic Measurement Theory* [16], which is an important part

Table 1.1. New classes of the numeral systems with irrational bases.

Title	Base	Mathematical formula
Bergman system (1957)	$\Phi = \frac{1+\sqrt{5}}{2}$, the root of the equation $x^2 - x - 1 = 0$	$A = \sum_i a_i \Phi^i, a_i \in \{0,1\}$ $\Phi^i = \Phi^{i-1} + \Phi^{i-2} = \Phi \times \Phi^{i-1}$
Codes of the golden p-proportions, $p = 0, 1, 2, 3, \ldots$ (Stakhov, 1984)	Φ_p is the root of the equation $x^{p+1} - x^p - 1 = 0,$ partial cases: $p = 0 \to \Phi_{p=0} = 2 \to$ binary system: $A = \sum_i a_i 2^i (a_i \in \{0,1\})$ $p = 1 \to \Phi_{p=1} = \Phi = \frac{1+\sqrt{5}}{2} \to$ Bergman system	$A = \sum_i a_i \Phi_p^i, a_i \in \{0,1\}$ $\Phi_p^i = \Phi_p^{i-1} + \Phi_p^{i-p-1} = \Phi_p \times \Phi_p^{i-1}$, partial cases: $p = 0 \to \Phi_{p=0} = 2 \to$ binary system $p = 1 \to \Phi_{p=1} = \Phi = \frac{1+\sqrt{5}}{2} \to$ Bergman system
Fibonacci p-codes $p = 0, 1, 2, 3, \ldots$ (Stakhov, 1972)	$\Phi_p = \lim_{p \to \infty} \frac{F_p(i)}{F_p(i-1)}$, the root of the equation $x^{p+1} - x^p - 1 = 0$	$N = a_n F_p(n) + a_{n-1} F_p(n-1) + \cdots + a_i F_p(i)$ $+ \cdots + a_n F_p(1), (a_i \in \{0,1\}),$ $F_p(i) = F_p(i-1) + F_p(i-p-1),$ partial cases: $p = 0 \to$ binary system, $p = 1 \to$ classical Fibonacci code
Ternary mirror-symmetrical representation (Stakhov, 2002)	$\Phi^2 = \frac{3+\sqrt{5}}{2}$	$N = \sum_i c_i (\Phi^2)^i, c_i \in \{\bar{1}, 0, 1\}$

of *Mathematics of Harmony* [6], generates not only a huge number of new, previously unknown measurement algorithms and positional numeral systems corresponding to them (see Table 1.1) but also an infinite number of new numerical sequences, in particular, a new class of recurrent numerical sequences, such as *Fibonacci p-numbers* ($p = 0, 1, 2, 3, \ldots$), including as special cases the "*binary*" *numbers* ($p = 0$) and the *classical Fibonacci numbers* ($p = 1$). These and other numerical sequences and the new number systems, generated by them (the *Fibonacci p-codes* [55] and the *codes of the golden p-proportions* [19]) should be the subject of study in the "golden" computer science, based on the new "*golden*" *paradigm* [99, 100].

1.2.4. Fibonacci *p*-codes

Just as the "binary" measurement algorithm "generates" the *binary numeral system* (1.2), the *Fibonacci measurement algorithms*, based on the *Fibonacci p-numbers* [16], "generate" new ways of the binary positional representation of natural numbers called *Fibonacci p-codes* [16]:

$$N = a_n F_p(n) + a_{n-1} F_p(n-1) + \cdots + a_i F_p(i) + \cdots + a_n F_p(1),$$

(1.5)

where $a_i \in \{0, 1\}$ is the binary numeral of the ith digit of the positional representation (1.5), n is the number of bits in the code (1.5), and $F_p(i)$ ($i = 1, 2, 3, \ldots, n$) is the weight of the ith digit, equal to the ith *Fibonacci p-number*. It is important to emphasize that the sum (1.5) includes an infinite number of different binary positional representations because each $p(p = 0, 1, 2, 3, \ldots)$ "generates" its own positional representation of the type (1.5).

As shown in Vol. II, the next special cases of the *Fibonacci p-codes* (1.5) are the following positional representations of natural numbers:

(1) The *classical binary method* of representation of natural numbers ($p = 0$):

$$N = a_n 2^{n-1} + a_{n-1} 2^{n-2} + \cdots + a_i 2^{i-1} + \cdots + a_1 2^0.$$ (1.6)

(2) The *classical Fibonacci code* ($p = 1$):

$$N = a_n F_n + a_{n-1} F_{n-1} + \cdots + a_i F_i + \cdots + a_1 F_1. \qquad (1.7)$$

(3) Finally, the *Euclidean definition* of the natural numbers $N(p = \infty)$, which underlies the classical *elementary number theory*, set forth in Euclid's *Elements* as follows:

$$N = \underbrace{1 + 1 + \cdots + 1}_{N}. \qquad (1.8)$$

It is necessary to emphasize once again the important role of the *Algorithmic Measurement Theory* [16] in creating the new mathematical concept, the *Fibonacci p-codes* (1.5). But the *Fibonacci p-codes*, given by the formula (1.5), are a generalization of the *binary system* (1.6) (for the case $p = 0$) and *classical Fibonacci code* (1.7) for the case ($p = 1$)!

Besides this, the *Euclidean definition* of the natural number N, given by the expression (1.8) (for the case $p = \infty$), is a particular case of the *Fibonacci p-codes* (1.5). It is clear that the concept of *Fibonacci p-codes* significantly expands our 59 number-theoretical ideas about the binary positional numeral systems, which directly follows from the *Algorithmic Measurement Theory* [16].

1.2.5. Codes of the "golden" p-proportions

But of even greater number-theoretical interest are the *codes of the golden p-proportions* introduced by Alexey Stakhov in the 1980 paper [59–61] and described in Stakhov's 1984 book [19]:

$$A = \sum_i a_i \Phi_p^i, \qquad (1.9)$$

where Φ_p is the "*golden*" *p-proportion*, the base of the *numeral system* (1.9), Φ_p^i is the weight of the ith digit, and $a_i \in \{0, 1\}$ is the *binary numeral* of the ith digit, $i = 0, \pm 1, \pm 2, \pm 3, \ldots$, $p = 0, 1, 2, 3, \ldots$.

Note that the expression (1.9) can be interpreted as the extension of the *Fibonacci p-codes* (1.5) to the area of real numbers, and from this point of view, the *codes of the golden p-proportions* are of special number-theoretical interest.

The codes of the *"golden" p-proportions* (1.9) include an infinite number of binary positional representations of real numbers (numeral systems) because every $p(p = 0, 1, 2, 3, \ldots)$ "generates" its own numeral system of the type (1.9). Note that for the case $p = 0$, the base $\Phi_p = \Phi_0 = 2$ and the numeral system (1.9) reduces to the *classical binary system*, which is the main principle of modern computer science:

$$A = \sum_i a_i 2^i. \tag{1.10}$$

Let's consider the case $p = 1$. For this case, the base of the numeral system (1.9) is the classical *golden proportion* $\Phi = \frac{1+\sqrt{5}}{2}$ and the numeral system (1.9) reduces to *Bergman's system* (1.1).

Let's now consider the extreme case $p \to \infty$. For this case, the base Φ_p tends to 1 which means that, in the limit, the formula (1.9) tends to the classical *Euclidean definition*, defined by (1.8), in which the "unit" ("monad") is the infinitely small segment $\Delta \to 0$.

Thus, we can consider the positional method of number representation, defined by (1.9), as a very broad generalization of the *Euclidean definition* (1.8) and also as the constructive definition of real numbers, based on the *binary system* (1.10) $(p = 0)$, and finally, as the *Bergman system* (1.1) $(p = 1)$ [54], the *first in the history of science numeral system with irrational base*.

1.2.6. The "golden" number theory

As it is known, the *elementary theory of numbers* begins with the *Euclidean definition* of natural number given by (1.8). Despite the seeming simplicity of the definition (1.8), it played a large role in the development of *elementary theory of numbers* and underlies many useful mathematical concepts, in particular, the concepts of *prime* and *composite* numbers, *multiplication, division, Euclidean algorithm*, and the concepts of *divisibility and comparisons*, which are among the basic concepts of *elementary theory numbers*.

In the article [74] published by Alexey Stakhov in the *Ukrainian Mathematical Journal* according to the recommendation of Academician Yuri Mitropolskiy, the *codes of the* "golden" *p-proportions* (1.9) and the *Bergman system* (1.1) are considered as a new approach to the geometric definition of real numbers.

As shown in Vol. II, the properties of numeral systems with irrational bases, given by (1.1) and (1.9), are unique from the number-theoretical point of view. Due to these numeral systems, our traditional ideas about numeral systems and also the relationships between rational and irrational numbers change. Thanks to the numeral systems (1.1) and (1.9), the *golden proportion* of ancient Greeks $\Phi = \frac{1+\sqrt{5}}{2}$ and its "homologues", the *golden p-proportions* Φ_p, unexpectedly entered into the "holy of holies" of mathematics, the theory of numbers, and it would be unwise to ignore this fact after the publication of the article [74].

In the *"golden" number theory* [74], the important number-theoretical results, concerning new properties of natural numbers, were obtained. In particular, it was proved that some natural number can always be represented as a finite sum of degrees of the *golden proportion* or the *golden p-proportion*, that is, the representation of natural numbers in the form

$$N = \sum_i a_i \Phi^i \tag{1.11}$$

or

$$N = \sum_i a_i \Phi_p^i \tag{1.12}$$

is always finite for the arbitrary natural number N.

But even more unexpected from a number-theoretical point of view are the so-called Z- and Z_p-properties of natural numbers [6, 74]. The essence of these properties is the fact that if in the expression (1.1), all the degrees of the *golden proportion* $\Phi^i (i = 0, \pm1, \pm2, \pm3, \ldots)$ are replaced with the corresponding classical *Fibonacci numbers* $F_i (i = 0, \pm1, \pm2, \pm3, \ldots)$ (according to the rule, $F_i \rightarrow \Phi^i$), and in the expression (1.9), all the degrees of the *golden p-proportions* Φ_p^i $(i = 0, \pm1, \pm2, \pm3, \ldots)$ are replaced with the corresponding *Fibonacci p-numbers* $F_p(i)(i = 0, \pm1, \pm2, \pm3, \ldots)$ (i.e., $F_p(i) \rightarrow \Phi_p^i$), then the sums, which arise here, are identically equal to 0, regardless of the original natural number N, that is,

$$\sum_i a_i F_i = 0, \tag{1.13}$$

$$\sum_i a_i F_p^i = 0; \tag{1.14}$$

moreover (and this is the most important thing!), *such a property is valid only for the natural numbers!*

Thus, more than two millennia after the beginning of the theoretical study of natural numbers (Euclid's *Elements*), the new number-theoretical properties of the natural numbers were found in the *"golden" number theory* [74], *which in itself is a scientific sensation.*

Of course, one can simply ignore the new mathematical properties of natural numbers, given by (1.11)–(1.14), but this is nothing but the "ostrich" policy. Probably, many sceptical mathematicians need to follow the example of the outstanding Ukrainian mathematician, academician of the Ukrainian and Russian Academies of Sciences, Yuri Mitropolskiy, who carefully studied this scientific direction and became its fan and defender (Fig. 1.2).

After careful study of Stakhov's scientific direction, academician Mitropolsky expressed his attitude to the *codes of the golden p-proportion* and the *Mathematics of Harmony* in the following words:

"Codes of the "golden" p-proportions. By using the concept of the "golden" p-proportion, Stakhov then introduces a new definition of a real number in the form (1.9), which is called the codes of the "golden" p-proportions. Stakhov shows that this concept, which is the development of the well-known Newtonian definition of a real number, can be used as the basis for a new theory of real numbers. Further, it shows that this result has important practical significance and may lead to the creation of fundamentally new computer arithmetic's and new computers, the "golden" computers. And Stakhov not only proclaims the idea of "Fibonacci computers", but also heads and organizes engineering projects for the creation of such computers at the Vinnitsa Polytechnic Institute (1977–1995).

Mitropolsky Yu.A.
(1917–2008)

- Doctor of Technical Sciences (1951)
- Professor (1953)
- Academician of the Academy of Sciences of Ukraine (1961)
- Honored Scientist of Ukraine (1967)
- Academician of the Russian Academy of Sciences (1984)
- Full member (academician) of the Taras Shevchenko Scientific Society (1992)
- Foreign Academic-Correspondent of the Academy of Sciences in Bologna (Italy, 1971)
- Hero of Socialist Labor and Hero of Ukraine

Fig. 1.2. Academician Yuri Mitropolskiy.

65 foreign patents for inventions in the field of the "golden" and Fibonacci computers, issued by state patent offices of the USA, Japan, England, France, Germany, Canada and other countries confirm the priority of Ukrainian science (and the priority of Prof. Stakhov) in this important computer area".

1.3. Fibonacci Microprocessors as a Prerequisite for the "Golden" Paradigm Shift in Computer Science

1.3.1. "Trojan Horse" of the binary system

Mathematics of Harmony overturned not only our ideas about the *elementary theory of numbers* but also radically changed the informational and arithmetical foundations of computer science. As it is well known, the *binary system* is the basis of modern computers. It was introduced to computer science in 1946 by John von Neumann together with his colleagues at the Princeton Institute for Advanced Study (IAS). The IAS is perhaps best known as the academic home of Albert Einstein, Hermann Weyl, John von Neumann and Kurt Gödel after their immigration to the United States.

The substantiation of *using of the binary system in electronic computers was one of the main "von Neumann principles."* At that time, this was an absolutely correct and weighted decision because the *binary system* most closely conforms to the *binary nature of electronic elements* and the requirements of the *Boolean logic.* In addition, it is necessary to take into consideration the fact that in that period, there were simply no other alternative numeral systems in science. *The choice was very small: a decimal system or a binary system. Preference was given to the binary system.*

However, together with the *binary system,* the "Trojan horse" was introduced into computer science in the form of *zero redundancy* of the *binary system.* The absence of redundancy means that all the binary code combinations within the binary system are "allowed", which makes it impossible (within the binary system) to detect any errors that inevitably (more or less likely) can occur in the elements of electronic systems under the influence of various external and internal factors (*radiation, electromagnetic effects, cosmic rays, interference in power supply tires,* etc.).

Thus, a far from optimistic conclusion follows from the foregoing. *The mankind is becoming hostage to the classical binary number system, which is the basis of modern microprocessors and information technology.* Therefore, **further development of microprocessors and information technology, based on the binary system, should be recognized as an unacceptable direction for many important applications of binary microprocessors and computers, in particular, the mission-critical applications.**

The *binary system* cannot be the informational and arithmetical basis for the specialized computer and measurement systems (*space systems, military systems, medical systems, complex technological objects,* etc.), as well as nano-electronic systems, where the problems of *reliability, noise immunity, stability,* and *survivability* of systems come to the fore.

Modern microprocessors and microcontrollers, based on the classical binary system, are unreliable from the informational point of view. A leading Russian expert in the field of specialized computer technology, academician Ya.A. Khetagurov, has long paid

Fig. 1.3. Academician Khetagurov.

attention to this serious shortcoming of the classical *binary system* (Fig. 1.3).

Academician Khetagurov in the 2009 article "Ensuring the national security of real-time systems" (BC/NW 2009; No. 2 (15)) expressed serious concern regarding the use of modern micropro-cessors and microcontrollers:

> *"The use of microprocessors, controllers and computational tools of foreign production for solving problems in real-time systems for military, administrative and financial purposes is fraught with great problems. This is a kind of "**Trojan horse**", whose role has only begun to emerge. Losses and harm from their use can significantly affect the national security of Russia".*

The above statement of the famous scientist is particularly relevant in connection with the frequent accidents that occur during the launch of Russian and US missiles. Possible causes of these accidents are failures in the digital control system, which has already been officially confirmed by the relevant Roscosmos commissions.

What is the output from this situation? Volume II is devoted to the study of the application of *Fibonacci codes* (1.5) and (1.7) and the *codes of the golden proportion* (1.9) for creating robust microprocessors. It is shown that these codes have a sufficiently high error-detecting ability (over 99%) and, based on them, *noise-resistant Fibonacci microprocessors* can be created for *mission-critical applications*. The theory of the *Fibonacci codes* and *codes of the golden proportions* continues to evolve intensively in the present time [6, 16, 19, 23, 53, 55, 56, 58–60, 65, 71, 97, 100]. Outstanding scientists and specialists are involved in the development of the Fibonacci circuit design, **and this process can be viewed as a "paradigm shift" in computer science for mission-critical applications.**

1.4. Sergey Abachiev: Mathematics of Harmony Through the Eyes of the Historian and Expert of Methodology of Science

1.4.1. Publication of Stakhov's article in the popular Soviet scientific journal *Technology to Youth*

The *codes of the golden proportion*, introduced in [19, 59], were always under the fixed attention of the wide scientific community. And this is confirmed by Stakhov's popular scientific article "Codes of the Golden Proportion, or numeral systems for future computers?" published in the popular Soviet scientific journal *Technique to Youth* (No. 7, 1985) in connection with the publication of Stakhov's 1984 book, *Codes of the Golden Proportion* [19] (Fig. 1.4).

1.4.2. Scientific work of Prof. Stakhov at the Dresden Technical University (April–May 1988)

In 1988, the administration of the Dresden Technical University (DTU) invited Alexey Stakhov for the 2-month work as a "Visiting Professor" of the Dresden Technical University at a special department for "visiting professors" named after Heinrich Barkhausen. Such a department was established at the Dresden Technical University in 1981 in honor of the 100th anniversary of the birth of Heinrich Barkhausen. This invitation was initiated by the famous German scientist, Academician Kempe, who at that time headed the Institute of Cybernetics and Informational Processes of the Academy of Sciences of the GDR. As it turned out later, Alexey Stakhov became the first Soviet professor, to be invited to work in this famous department.

Fig. 1.4. Journal *Technology to Youth*, No. 7, 1985.

But who was Heinrich Barkhausen? Heinrich Georg Barkhausen (1881–1956) was an outstanding German scientist in the field of electronic physics and electrical engineering, the founder of modern low-current technology and the latest electronics. Since 1911, Heinrich Barkhausen worked as a Professor at the Higher Technical School in Dresden. During 1917–1918, Barkhausen, independent of other researchers, created the theory of tube generator. In 1919, he discovered the phenomenon of discontinuity in ferromagnetism, which was called the *Barkhausen effect*. In 1928, he was awarded the Heinrich Hertz Medal: in 1949, he received the National Prize of the GDR.

The work of Alexey Stakhov at the Dresden Technical University was quite intense. Firstly, he delivered the course of lectures *Fibonacci Numbers and Computers* for the undergraduate and graduate students of the Dresden Technical University. Secondly, he delivered the lectures and speeches at various universities and research institutions of the GDR, in particular, at the Karl-Marx-Stadt Technical University and the Institute of Cybernetics and

Informational Processes of the Academy of Sciences of the GDR (Berlin).

1.4.3. Speech at the Karl-Marx-Stadt Technical University

Stakhov's speech at the seminar of the Computer Science Department of the Karl-Marx-Stadt Technical University was one of the most memorable events of Stakhov's stay in the GDR. According to the results of this speech, the Director of the Department, Prof. Postoff, wrote the following review on Stakhov's speech:

"22.04.1988 in the framework of the seminar of our department Prof. Stakhov A.P. from Vinnitsa made a speech 'The Fibonacci numbers in computer science' in German. In accordance with the requirements of the topic, the speech addressed to a wide range of issues, related to the golden section, ranging from the appearance of golden sections and Fibonacci numbers in art and biological forms, geometry, arithmetic, and ending with their use to create simple, reliable, self-correcting codes and measurement algorithms in computer science.

The topic of the speech is of interest to a wide range of listeners, but how to present this material depends largely from the speaker; Prof. Stakhov perfectly used all the possibilities and made a deep impression.

Since the Fibonacci numbers play an important role in explaining the most diverse phenomena in nature, art, science and technology, it is not surprising that different scientists independently came to these dependencies. In addition to Prof. Stakhov, the Englishman Turing, the founder of theoretical computer science, also dealt with these problems.

On the other hand, this topic is in no way generally accessible from a scientific point of view, because scientists, studying this topic, must be able to unite together the most essential, unusually extensive and heterogeneous knowledge of unusually extensive and heterogeneous fragments of human culture. Such scientists rarely appear, and their cooperation is necessary and useful, as in no other field.

Doctor of Technical Sciences, Professor Postoff,

Director of the Computer Science Department, Karl-Marx-Stadt Technical University"

1.4.4. Publication of the interview in the newspaper "Pravda"

In connection with the successful completion of the 2-month work of Alexey Stakhov at the Dresden Technical University, the administration of DTU awarded of Stakhov with the **Memorial Medal of Heinrich Barkhausen** as "Visiting Professor". The Soviet media drew special attention to the successful results of Stakhov's work in the GDR.

The most significant event was the publication of Stakhov's interview in the newspaper *Pravda*, the central newspaper of the Soviet Union (November 19, 1988). This interview could not pass by party organs, government circles and the scientific community of the USSR.

This interview by Stakhov aroused great interest of the Soviet scientific community. About a 100 enthusiastic letters came to Alexey Stakhov from all over the Soviet Union. The letter of Professor Sergey Abachiev (Moscow), PhD in the history of science and technology, was the most interesting (Fig. 1.5).

The area of Abachiev's research interests is extremely broad: traditional formal logic and evolutionary theory of cognition, epistemological aspects of labor and technology, modern scientific-methodological and spiritual self-cognition of philosophy and so on.

Since 1978, Professor Sergey Abachiev became and remains a convinced supporter of Stakhov's *Mathematics of Harmony*. He took

- PhD in the history of science and technology

- Professor of the Department of Philosophy and Worldview Security at the Institute of Public Administration, Law and Innovation Technologies (Moscow)

Fig. 1.5. Sergey Abachiev.

an active participation in the discussion of this scientific direction on the website of the Academy of Trinitarism (Moscow, Russia) and he published a series of in-depth articles, in which he analyzed the current state and prospects of the development of *Mathematics of Harmony*.

Unlike other scientists, who took part in the discussion of Stakhov's *Mathematics of Harmony*, Prof. Abachiev in his critical articles focuses on Stakhov's main scientific achievements in computer science: the *codes of the golden proportions*, the *Fibonacci p-codes*, the *Fibonacci arithmetic*, and the *concept of Fibonacci computers* as a new trend in computer technology. By analyzing the reasons of the difficult historical fate of *Mathematics of Harmony*, Prof. Abachiev wrote the following in one of his papers:

"A scientific direction, developing by Stakhov and his students for more than forty years..., seems to me one of the rare examples of the organic joining of the fundamental and applied sciences...I see here direct analogies, first of all, with the history of technology... The difficult historical fate of Stakhov's scientific direction was determined by its historical randomness, which is not connected with the basic laws of the development of science and technology. The discovery by the 12-year-old child prodigy Georg Bergman of the "golden" numeral system with irrational base was in no way, predetermined by such laws. Stakhov's scientific achievements possibly could be got many decades earlier, but possibly could not be got to this day. But already in 1957, when Bergman's discovery was actually done, the flywheel of the digital information technology industry was unwound on the basis of Shannon's theory of information and von Neumann's binary code. And this flywheel was fully promoted to the beginning of the 70s, when Stakhov for the first time appreciated the "golden" numeral system as the arithmetic fundamental principle of digital informational technology.

Von Neumann's choice of the binary code with all its shortages compared to redundant codes of the golden proportion should not be considered as historically unsuccessful and erroneous. In the late 40s the binary code simply had no alternatives. In principle, Bergman's amateur discovery, dated 1957, could be made by someone else half a century earlier.

If the first "golden" numeral system would come to the attention of Hartley, Shannon and von Neumann, the history of digital information technology could begin immediately with the golden proportion codes. But the real history of world science and technology decreed differently. Alexey Stakhov became the first inventor and professional designer of this amateur discovery in the conditions of the untwisted flywheel of information technology, based on the binary code... But the real history of world science and technology decreed differently. Alexey Stakhov became the first inventor and professional designer of this amateur discovery in the conditions of the untwisted flywheel of information technology, based on the binary code... Taught by the bitter experience of past persecutions of genetics and cybernetics, the Soviet state quickly realized that domestic science can get strategically breakthrough positions in the all-determining direction of scientific and technological progress. The evidence of this was the unprecedented patenting of the first information technologies by A. P. Stakhov on a qualitatively new arithmetical fundamental principle in the USSR, in the West and in Japan.... However, such technologies objectively could not then quickly oust the completely dominant technologies based on binary code. In any case, their expansion would be a purely phased process, lasting many decades.

And in the 80s years this natural process in our then still united country began to be carried out in the relatively narrow field of onboard electronics of military aircraft and spacecraft, where the economic criteria for the effectiveness of technology move away into the background compared to functional.

With normal development, by now it would allow for Russia and Ukraine to be World "legislators" and manufacturers of at least uniquely reliable avionics. But the disastrous finale of "perestroika" 1985–1991 already in the initial phase this process of achievement by our country of the leading world positions in the field of technical cybernetics and information technologies was cut short.

To the present date, several key areas of the development of information technologies urgently require the usage of the redundant, especially reliable and noise-resistant codes of the golden proportions....

This direction of information technology development in general seems to be one of those several innovative directions that should be the subject of special state attention and special state care. It can and should become one of the key in the post-Soviet scientific, technical and economic reintegration of Russia and Ukraine, along with the revival of cooperation in the aerospace industry."

A history of the implementation of the "golden" paradigm shift into modern science is very reminiscent of the history of the recognition of Lobachevsky's geometry in mathematics of the 19th century. After a sharply negative assessment of Lobachevsky's mathematical discovery, made by the leading Russian mathematician academician Ostrogradsky, the mathematical community of Russia turned away from Lobachevsky's geometry, and only thanks to the positive assessment of the new geometry, made by the outstanding German mathematician *Johann Carl Friedrich Gauss* (1777–1855), the world mathematical community were forced to recognize Lobachevsky's geometry as the greatest mathematical discovery of the 19th century. Unfortunately, this well-deserved recognition came to Lobachevsky only 10 years after his death.

Chapter 2

The "Golden" Hyperbolic Functions as the "Golden" Paradigm Shift to the "Golden" Non-Euclidean Geometry

2.1. The Concept of "Elementary Functions"

Among the huge number of different mathematical functions, we can distinguish their special class called the *elementary functions*. The significance of these functions for science consists in the fact that they express some of the stable relationships that occur in the world around us. Examples of the *elementary functions* are as follows: *power, exponential, logarithmic, trigonometric, hyperbolic functions*, etc.

The question arises: What is the sense of the adjective *elementary* in this definition? Most of the *elementary functions* are studied into school program. Hence, there is a traditional desire to interpret the *elementary functions* as very "*simple*", "*school*" functions.

However, in English, the word *elementary* means *original, primary, fundamental*. This word has the same meaning when we say Euclid's *Elements* in geometry or *elementary particles* in theoretical physics. Therefore, the phrase *elementary functions* should be understood not as "*simple*", "*school*" functions, but as *primary, fundamental functions*, that is, the functions that are widely manifested themselves in Nature. It is these functions that link mathematics with theoretical natural sciences. Therefore, the introduction of new types of *elementary functions* that underlie

25

some natural phenomena and processes should be considered as an important scientific achievement and as a new "shift of paradigm" in elementary functions.

Let's note that the adjective "elementary" has a similar meaning when we speak about *elementary mathematics*. *Elementary mathematics* is not "school mathematics", as is commonly believed; it is the initial part of mathematics, which contains the original, fundamental concepts of mathematics, without which "higher mathematics" simply could not exist.

Let's try to exclude from mathematics such "elementary concepts" as *natural* and *irrational* numbers, the *Pythagorean theorem*, *fundamental mathematical constants*, the *constant of* π and the *Euler constant of e*, and then let's consider what remains as "higher mathematics"?

Therefore, obtaining new mathematical results and proving new theorems in the field of the *elementary mathematics*, in particular, in the *theory of the elementary functions* is of great interest both for *elementary* and for *higher* mathematics. This chapter is devoted to the discussion of the new results in the field of *elementary functions* called *Fibonacci and Lucas hyperbolic functions* [64, 75] and their applications in the theoretical natural sciences (in particular, in the *new geometric theory of phyllotaxis* [42]).

2.2.   Conic Sections and Hyperbola

2.2.1.   Apollonius' "conic sections"

One of the main goals of this volume is the study of a new class of hyperbolic functions, that is, the *"golden" hyperbolic functions* called the *Fibonacci and Lucas hyperbolic functions* [64, 75]. This is a new kind of hyperbolic functions related to *elementary functions*. Their unusuality consists in the fact that, unlike the classical hyperbolic functions with the base e (the Euler constant), the base of the "golden" hyperbolic functions is the "golden proportion". Let's start with the source of the notion of *hyperbolic functions*. In this regard, it is appropriate to recall one more great mathematical discovery of antiquity: the *conic sections*. The author of the *conic sections* was

Apollonius of Perga, one of the greatest mathematicians of antiquity, who lived in the 3rd–early 2nd centuries BC. Like Pythagoras, he went to Alexandria, the most important center of Hellenistic culture, where he studied mathematics; his teachers were Euclid's students.

Apollonius wrote a number of works on mathematics and optics, but his essay *Conic Section* is considered the most famous. This fundamental mathematical work consisted of eight books; only the first four books came to us in the original form, the next three books came to us in the form of Arabic translation, the last book was lost. In his first book, he examines the various flat sections of one and the same circular cone (the "conic sections"); its cavities extend on both sides of its top (Fig. 2.1).

Depending on whether the plane intersects only one plane of the cone, both its planes or it is parallel to one of the generators of the cone, Apollonius gets the following remarkable curves: (1) parabola, (2) circle and ellipse, (3) hyperbola.

Based on some geometric properties of the *conic sections*, Apollonius chose appropriate names for them. The Greek word *parabole* means *appendix*, the Greek word *elleipsis* means *"flaw"*, the Greek word *hyperbole* means *"excess"*, *"exaggeration"*.

In the ancient times, the use of conic sections in science and practice was bounded. The *conic section* began to play a major role in

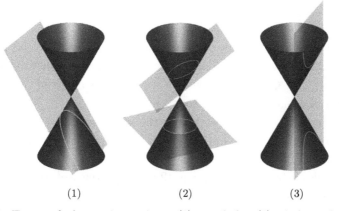

| (1) | (2) | (3) |

Fig. 2.1. Types of the conic sections: (1) parabola, (2) circle and ellipse, (3) hyperbola.

science and technology in the modern times when Galileo established that a freely thrown body or projectile, fired from a weapon, moves along a *parabola*.

But a special interest in the *conic section* appeared after Kepler formulated the laws of planet motion. According to the first Kepler law, each planet moves along an *ellipse* and in its focus, the Sun is located. Later, it was found that some comets move in *ellipses*, others in *parabolas* and *hyperbolas*.

The *conic section* is widely used in science and technology. A parabolic mirror has the property that all incident rays parallel to its axis converge at one point (the *focus*). This is used in most reflecting telescopes, where parabolic mirrors are used, as well as in radar antennas and special microphones with parabolic reflectors. A beam of parallel rays emanates from a light source placed at the focus of the parabolic reflector. Therefore, the parabolic mirrors are used in high-powered searchlights and car headlights.

2.2.2. Hyperbola

In the future, an important role will be played by the graph of inverse proportionality, that is, a curve, the equation of which has the following form:

$$y = \frac{a}{x} \quad \text{or} \quad xy = a. \tag{2.1}$$

This curve is called *hyperbole*.

Hyperbola is a graph of many important physical relationships, for example, Boyle's law (connecting the pressure and volume of an ideal gas) and Ohm's law, which specifies an electric current as a function of resistance at the constant voltage.

To describe the properties of the *hyperbola* (2.1), we use the reasoning given in Refs. [1, 140]. The graph of the *hyperbola* is depicted in Fig. 2.2.

As follows from the formulas (2.1) and the graph in Fig. 2.2, the *hyperbola* consists of two branches, which for $a > 0$ are located in the first quadrant of the coordinate system (x and y are positive) and in the third quadrant (x and y are negative). Geometrically,

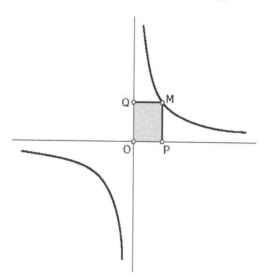

Fig. 2.2. Hyperbola and its coordinate rectangle.

the *hyperbola* approaches arbitrary close to the coordinate axes, never intersecting them. This means that the coordinate axes are asymptotes of the *hyperbola*.

Note that the *hyperbola equation* $xy = a$ has a simple geometric interpretation: the area of the rectangle $MQOP$, bounded by the *coordinate axes* and the lines, drawn through any point M of the hyperbola, parallel to the coordinate axes (Fig. 2.2), is equal to a. Let's name the rectangle $MQOP$, whose area is equal to a, by the *coordinate rectangle* of the point M; then we can give the following geometric definition of the *hyperbola*:

> *"A hyperbola is a geometric place of the points that lie in the first and third quadrants of the coordinate system, whose coordinate rectangles have a constant area."*

It is easy to see that the origin of coordinates O is the *center of symmetry* of the *hyperbola*, that is, the branches of the hyperbola are symmetrical to each other relative to the origin of coordinates O. The hyperbola also has *symmetry axes* that are the bisectors of the angle coordinates aa and bb (Fig. 2.3).

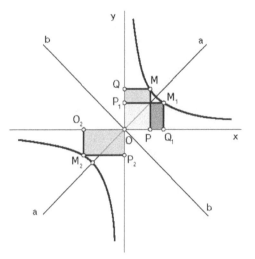

Fig. 2.3. Hyperbola axes aa and bb.

The center of symmetry O and the axes of symmetry aa and bb are often called simply the *center* and *axes of the hyperbola*; the points A and B, in which the hyperbola intersects with the axis aa are called the *vertices of the hyperbola*.

2.3. Hyperbolic Rotation

2.3.1. Compression to a point

Geometry often uses a geometric transformation called *compression to a point* or *homothety*. The compression to the point O, called a *compression center*, with the *compression coefficient* k, consists in the following. Each point A of the plane goes to the point A', which lies on the beam OA, moreover $OA' = kOA$. If the compression coefficient $k > 1$, then $OA' > OA$; in this case, the transformation should be called the *extension from O*.

It has been proven in [140] that the *compression to a point* has the following properties:

(1) Each geometric figure F for the *compression to a point* goes into the figure F' similar to the original. If $k < 1$, then the figure decreases, and if $k > 1$, then the figure increases.

(2) When a straight line is compressed to a point, it remains a straight line; parallel straight lines remain parallel.
(3) When a circle is compressed to a point, it turns into a circle.
(4) All segments of the plane under compression to a point decrease (or increase) in a constant ratio k.
(5) The areas of all geometric figures under the compression to a point decrease (or increase) in the constant ratio equal to k^2.

2.3.2. Compression to straight line

Often, another transformation is used in geometry called *the compression to a straight line* [140]. The compression to a straight line, which is called a compression axis with the compression coefficient k consists in the fact that each point A of the plane passes to the point A', which lies on the beam PA, perpendicular to the compression axis, moreover the following relationship has a place: $PA' = kPA$. Note that if the "compression ratio" $k > 1$, then $PA' > PA$; in this case, the transformation should be called *extension from the compression axis*. The following geometric properties of the *compression* are established in [140]:

(1) When compressing to a straight line, every straight line turns into a straight line.
(2) When compressing to straight lines, parallel straight lines remain parallel.
(3) During compression to a straight line, the ratio of segments, lying on one straight line, is preserved.
(4) When compressing to a straight line, all geometric figures change in a constant ratio (equal to the compression coefficient k).

2.3.3. Hyperbolic rotation

Let us now consider the hyperbola $xy = a$ (Fig. 2.2). Let's make the compression of a plane to the axis x with the *compression coefficient* k. In this case, the hyperbola $xy = a$ will go to the hyperbola $xy = ak$ because the abscissa x will remain unchanged, and the ordinate y will be replaced by yk.

Then we make another compression to the axis y with the coefficient $\frac{1}{k}$. In this case, the hyperbola $xy = ak$ goes into the hyperbola $xy = \frac{ak}{k} = a$: the ordinate y of any point at new compression to the axis does not change, and the abscissa x goes into $\frac{x}{k}$. Thus, we see that the successive compression of the plane to the axis x with the coefficient k and then to the axis y with the coefficient $\frac{1}{k}$ passes the hyperbola $xy = a$ into itself.

The succession of these two compressions of the plane to the axes x and y forms an important geometric transformation, which is called the *hyperbolic rotation*. The name of the *hyperbolic rotation* reflects the fact that with such a transformation, all points of the hyperbola *slide along a curve*, that is, the *hyperbola* as if *turns*.

The following properties of the *hyperbolic rotation* directly follow from the above-considered properties of the *compression* [140]:

(1) With the *hyperbolic rotation*, every straight line turns into a straight line.
(2) With the *hyperbolic rotation*, the axes of coordinates (asymptotes of the hyperbola) go into themselves.
(3) With the "hyperbolic rotation", the parallel lines remain parallel.
(4) With the "hyperbolic rotation", the areas of the geometric figures are preserved.

It is very important to emphasize that by choosing the appropriate value of the *compression coefficient* k with the help of the *hyperbolic rotation*, each point of the hyperbola can be transferred to some other.

2.4. Trigonometric Functions

2.4.1. Geometric definition

Let's consider *trigonometric functions*, the simplest class of the *elementary functions*, well known to us from secondary school. Usually, this class of "elementary functions" includes the following: *sine, cosine, tangent, cotangent, secant*, and *cosecant*; the last pair of functions (*secant* and *cosecant*) is currently relatively rarely used.

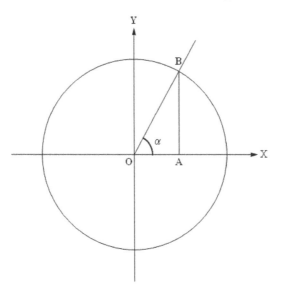

Fig. 2.4. Geometric definition of trigonometric functions.

Usually, the *trigonometric functions* are defined geometrically by using a circle (Fig. 2.4). For this purpose, we will measure the angles, which are formed in a circle by rotating the beam *OB* in the direction from the abscissa axis *OX*. In this case, the counterclockwise direction is considered as positive, and the clockwise direction is considered as negative. The abscissa of the point *B* is denoted by x_B, and the ordinate is denoted by y_B (Fig. 2.4).

Then we can give the following geometric definition of the trigonometric functions:

(1) *Sine* is the ratio: $\sin \alpha = y_B/R$;
(2) *Cosine* is a ratio: $\cos \alpha = x_B/R$;
(3) *Tangent* is defined as: $\operatorname{tg} \alpha = \sin \alpha / \cos \alpha$;
(4) *Cotangent* is defined as: $\operatorname{ctg} \alpha = \cos \alpha / \sin \alpha$;
(5) *Secant* is defined as: $\sec \alpha = 1 / \cos \alpha$;
(6) *Cosecant* is defined as: $\operatorname{cosec} \alpha = 1 / \sin \alpha$.

It is clear that the values of trigonometric functions do not depend on the radius R of the circle due to the properties of this

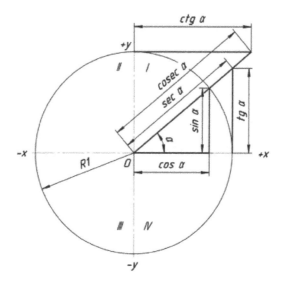

Fig. 2.5. Geometric definition of the trigonometric functions for the unit circle.

geometric figure. Often, the radius is taken equal to the value of the unit segment, then the *sine* is simply equal to the ordinate y_B and cosine is equal to the abscess x_B. Fig. 2.5 shows the values of the above trigonometric functions for the unit circle.

2.4.2. The simplest identities for the trigonometric functions

Since the *sine* and *cosine* are, respectively, the *ordinate* and the *abscissa* of the point, corresponding to the angle α on the unit circle (Fig. 2.5), we can get the first remarkable identity for trigonometric functions by using the *Pythagorean Theorem*:

$$\cos^2 \alpha + \sin^2 \alpha = 1. \tag{2.2}$$

If we divide all the terms of the identity (2.2) first by the *square of the cosine* and then by the *square of the sine*, respectively, we get the two well-known identities for the trigonometric functions:

$$1 + \operatorname{tg}^2 \alpha = \frac{1}{\cos^2 \alpha}, \quad 1 + \operatorname{ctg}^2 \alpha = \frac{1}{\sin^2 \alpha}. \tag{2.3}$$

There are a number of the identities that define the important *parity property* for the trigonometric functions:

$$\sin(-\alpha) = -\sin\alpha, \quad \cos(-\alpha) = \cos\alpha,$$
$$\mathrm{tg}(-\alpha) = -\mathrm{tg}\,\alpha, \quad \mathrm{ctg}(-\alpha) = \mathrm{ctg}\,\alpha, \qquad (2.4)$$
$$\sec(-\alpha) = \sec\alpha, \quad \mathrm{cosec}(-\alpha) = -\mathrm{cosec}\,\alpha.$$

It follows from (2.4) that the *cosine* and *secant* are the *even* functions, and the other four functions are the *odd* functions.

2.5. Geometric Analogies Between Trigonometric and Hyperbolic Functions and Basic Identities for Hyperbolic Functions

2.5.1. The hyperbola equation, related to the axes

The reasoning here is taken from Shervatov's wonderful booklet *Hyperbolic Functions* [140]. As before, the coordinates of the hyperbola point M in the coordinate system, the axes of which coincide with the asymptotes, will be x and y, so the hyperbola equation will look like $xy = a$ (Fig. 2.6).

Let's consider the axes aa and bb of the *hyperbola* as new coordinate axes. The coordinates of the point M in the new

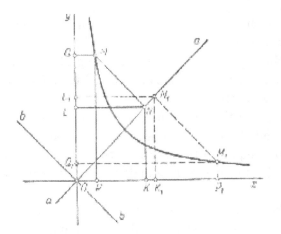

Fig. 2.6. The derivation of the hyperbola equation related to the axes.

coordinate system are denoted by X and Y. It is easy to show that the old coordinates x and y can be expressed in terms of the new coordinates X and Y as follows:

$$x = (X - Y)\frac{\sqrt{2}}{2}, \quad y = (X + Y)\frac{\sqrt{2}}{2}.$$

If we now substitute the obtained values of x, y into the equation $xy = a$, then, after carrying out the corresponding transformations, we obtain the equation:

$$X^2 - Y^2 = 2a. \tag{2.5}$$

This is the *hyperbola equation* related to the axes. For the case $a = \frac{1}{2}$, we get from (2.5) the following equation of the *unit hyperbola*:

$$X^2 - Y^2 = 1. \tag{2.6}$$

2.5.2. Geometric analogies between trigonometric and hyperbolic functions

In Ref. [140], a concept of the *hyperbola radius*, which is a segment of diameter that goes from the *center of the hyperbola* to the point of intersection of the diameter with the hyperbola (i.e., the radii of the hyperbola are defined similar to the radii of the circle), is also introduced.

By using the above geometric approach to the definition of trigonometric functions (Figs. 2.4 and 2.5), we will set forth the geometric theory of trigonometric and hyperbolic functions based on the analogy between the *unit circle* (Fig. 2.7(a)) and the *unit hyperbola* (Fig. 2.7(b)).

Let's consider the *unit circle* (Fig. 2.7(a)): $X^2 + Y^2 = 1$. The angle α (in radial measure) between radii OA and OM of the circle is the number, equal to the length of the arc AM or equal to twice the area of the sector OAM, bounded by these radii and the arc of the circle.

From the point M, we now drop the perpendicular MP to the diameter OA; in the point A, we draw the tangent to the circle to the intersection with the diameter OM in the point N. The segment

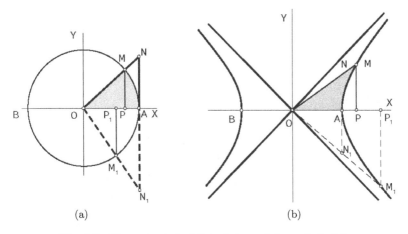

Fig. 2.7. The unit circle (a) and the unit hyperbole (b).

PM of the perpendicular is the line of *sine*, the segment OP of the diameter is the line of *cosine* and the segment AN is the line of *tangent*. The lengths of the segments PM, OP and AN are equal to the *sine*, *cosine*, and *tangent* of the angle α, respectively:

$$PM = \sin\alpha, \quad OP = \cos\alpha, \quad AN = \operatorname{tg}\alpha. \qquad (2.7)$$

Now, by analogy, we will introduce the concepts of the *hyperbolic sine*: the *hyperbolic cosine* and the *hyperbolic tangent*. To do this, let's consider the *unit hyperbole* (Fig. 2.7(b)): $X^2 - Y^2 = 1$. Then, by analogy with the angle α between the radii OA and OM (Fig. 2.7(a)), the hyperbolic angle t between the two radii OA and OM of the hyperbola (Fig. 2.7(b)) will be called the number, equal to twice the area of the sector, bounded by these radii and arc of the *hyperbola*.

Now, let's drop from the point M of the hyperbola in Fig. 2.7(b) the perpendicular MP to the diameter OA, which is the *axis of symmetry*, intersecting the hyperbola at vertex A; after this, we will draw the tangent to the hyperbola to the intersection at the point N of the diameter OM.

The segment PM of the perpendicular is called the line of the *hyperbolic sine*, the segment of the diameter OP is called the line of the *hyperbolic cosine* and the segment AN is called the line of the *hyperbolic tangent.* The lengths of the segments *PM, OP,* and *AN* are called the *hyperbolic sine, hyperbolic cosine,* and *hyperbolic tangent* of the angle *t,* respectively:

$$PM = \operatorname{sh} t, \quad OP = \operatorname{ch} t, \quad AN = \operatorname{th} t. \qquad (2.8)$$

It is known that the *trigonometric functions* (2.7) change periodically with the period 2π. In contrast, the *hyperbolic functions* (2.8) are *not periodic.* As it follows from Fig. 2.7(b), the *hyperbolic angle t* can vary from 0 to ∞. From the geometric definition of the *hyperbolic functions* (Fig. 2.7(b)), it follows that when the hyperbolic angle changes from 0 to ∞, then $\operatorname{sh} t$ changes from 0 to ∞, $\operatorname{ch} t$ changes from 1 to ∞ and $\operatorname{th} t$ changes from 0 to 1. It is easy to show that these functions have graphs, the form of which is shown in Fig. 2.8.

Let's now derive the main dependences between trigonometric circular and hyperbolic functions.

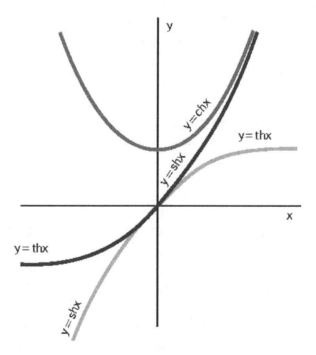

Fig. 2.8. Graphs of the hyperbolic functions.

From the similarity of the triangles OMP and ONA (Fig. 2.7(a)), it follows that: $\frac{AN}{OA} = \frac{PM}{OP}$. But $\frac{AN}{OA} = \operatorname{tg}\alpha$ (because $OA = 1$), and $\frac{PM}{OP} = \frac{\sin\alpha}{\cos\alpha}$. Thus, we get $\operatorname{tg}\alpha = \frac{\sin\alpha}{\cos\alpha}$.

Further, the coordinates of the point M of the circle are $OP = X$, $PM = Y$. But then from the equation of the *unit circle* (Fig. 2.7(a)), it follows that $OP^2 + PM^2 = 1$ or $\cos^2\alpha + \sin^2\alpha = 1$.

By dividing both parts of the obtained identity $\cos^2\alpha + \sin^2\alpha = 1$ first by $\cos^2\alpha$, and then by $\sin^2\alpha$, we get two remarkable formulas for the trigonometric functions given by (2.3).

Now, we will derive similar properties for the hyperbolic functions (2.8). From the similarity of the triangles OMP and ONA (Fig. 2.7(b)), it follows that: $\frac{AN}{OA} = \frac{PM}{OP}$. But $\frac{AN}{OA} = \operatorname{th}t$ (because $OA = 1$), and $\frac{PM}{OP} = \frac{\operatorname{sh}t}{\operatorname{ch}t}$. Thus, we get $\operatorname{th}t = \frac{\operatorname{sh}t}{\operatorname{ch}t}$.

Further, the coordinates of the point M of the hyperbola are $OP = X, PM = Y$.

But then from the equation of the *unit hyperbola* (2.5), it follows that

$$OP^2 - PM^2 = 1 \quad \text{or} \quad \operatorname{ch}^2 t - \operatorname{sh}^2 t = 1. \tag{2.9}$$

By dividing both parts of the identity (2.9) first by $\operatorname{ch}^2 t$ and then by $\operatorname{sh}^2 t$, we obtain the two remarkable formulas for the hyperbolic functions, which are analogous to the formulas (2.3):

$$1 - \operatorname{th}^2 t = \frac{1}{\operatorname{ch}^2 t}, \quad \operatorname{ch}^2 t - 1 = \frac{1}{\operatorname{sh}^2 t}. \tag{2.10}$$

Similarly, by analogy, other identities for the hyperbolic functions (2.8) can be proved in particular:

$$\operatorname{sh}(t + u) = \operatorname{sh}t\operatorname{ch}u + \operatorname{ch}t\operatorname{sh}u,$$
$$\operatorname{ch}(t + u) = \operatorname{ch}t\operatorname{ch}u + \operatorname{sh}t\operatorname{sh}u,$$
$$\operatorname{sh}2t = 2\operatorname{sh}t\operatorname{ch}t,$$
$$\operatorname{ch}2t = \operatorname{ch}^2 t + \operatorname{sh}^2 t.$$

2.5.3. Analytic expressions for hyperbolic functions

Unlike the trigonometric functions, the hyperbolic functions can be expressed analytically [140]. Their analytical expressions are as follows:

Hyperbolic sine and cosine:

$$\mathrm{sh}\,x = \frac{e^x - e^{-x}}{2}, \quad \mathrm{ch}\,x = \frac{e^x + e^{-x}}{2}, \tag{2.11}$$

Hyperbolic tangent and cotangent:

$$\mathrm{th}\,x = \frac{\mathrm{sh}\,x}{\mathrm{ch}\,x}, \quad \mathrm{cth}\,x = \frac{\mathrm{ch}\,x}{\mathrm{sh}\,x}. \tag{2.12}$$

The analytical expressions (2.11) allow us to prove a number of important identities for the hyperbolic functions, some of which we established earlier geometrically, in particular:

$$\mathrm{ch}^2\,x - \mathrm{sh}^2\,x = \left(\frac{e^x + e^{-x}}{2}\right)^2 - \left(\frac{e^x - e^{-x}}{2}\right)^2$$

$$= \frac{e^{2x} + 2 + e^{-2x}}{4} - \frac{e^{2x} - 2 + e^{-2x}}{4} = 1, \tag{2.13}$$

$$\mathrm{ch}^2\,x + \mathrm{sh}^2\,x = \left(\frac{e^x + e^{-x}}{2}\right)^2 + \left(\frac{e^x - e^{-x}}{2}\right)^2$$

$$= \frac{e^{2x} + 2 + e^{-2x}}{4} + \frac{e^{2x} - 2 + e^{-2x}}{4}$$

$$= \mathrm{sh}(2x). \tag{2.14}$$

By using (2.11), and (2.12), we can prove the following identities for the hyperbolic functions:

Parity property:

$$\mathrm{sh}(-x) = -\mathrm{sh}\,x, \quad \mathrm{ch}(-x) = \mathrm{ch}\,x, \quad \mathrm{th}(-x) = -\mathrm{th}\,x, \tag{2.15}$$

Formulas for the addition:

$$\text{sh}(x \pm y) = \text{sh}\,x\,\text{ch}\,y \pm \text{sh}\,y\,\text{ch}\,x, \text{ch}(x \pm y)$$
$$= \text{ch}\,x\,\text{ch}\,y \pm \text{sh}\,y\,\text{sh}\,x, \text{th}(x \pm y)$$
$$= \frac{\text{th}\,x \pm \text{th}\,y}{1 \pm \text{th}\,x\,\text{th}\,y}$$

Formulas for the dual angle:

$$\text{sh}(2x) = 2\,\text{ch}\,x\,\text{sh}\,x = \frac{2\,\text{th}\,x}{1 - \text{th}^2 x},$$
$$\text{ch}(2x) = \text{ch}^2 x + \text{sh}^2 x = 2\,\text{ch}^2 x - 1,$$

Formulas for the product:

$$\text{sh}\,x \cdot \text{ch}\,y = \frac{\text{sh}(x+y) + \text{sh}(x-y)}{2},$$
$$\text{sh}\,x \cdot \text{sh}\,y = \frac{\text{ch}(x+y) - \text{ch}(x-y)}{2},$$
$$\text{ch}\,x \cdot \text{ch}\,y = \frac{\text{ch}(x+y) + \text{ch}(x-y)}{2}.$$

Formulas for the summations:

$$\text{sh}\,x \pm \text{sh}\,y = 2\,\text{sh}\,\frac{x \pm y}{2} \cdot \text{ch}\,\frac{x \mp y}{2},$$
$$\text{ch}\,x + \text{ch}\,y = 2\,\text{ch}\,\frac{x + y}{2} \cdot \text{ch}\,\frac{x - y}{2},$$
$$\text{ch}\,x - \text{ch}\,y = 2\,\text{sh}\,\frac{x + y}{2} \cdot \text{sh}\,\frac{x - y}{2}.$$

Formulas for derivatives:

$$(\text{sh}\,x)' = \text{ch}\,x, \quad (\text{ch}\,x)' = \text{ch}\,x, \quad (\text{th}\,x)' = \frac{1}{\text{ch}^2 x}.$$

Formulas for the integrals:

$$\int \mathrm{sh}(x)dx = \mathrm{ch}(x) + C, \quad \int \mathrm{ch}(x)dx = \mathrm{sh}(x) + C,$$

$$\int \mathrm{th}(x)dx = \ln[\mathrm{ch}(x)] + C, \quad \int \frac{1}{\mathrm{ch}^2(x)}\, dx = \mathrm{th}(x) + C,$$

$$\int \frac{1}{\mathrm{sh}^2(x)}dx = -\mathrm{cth}(x) + C.$$

2.5.4. Applications of the hyperbolic functions in geometry

It is believed that the hyperbolic functions were introduced by the Italian mathematician Vincenzo Riccati in 1757. He obtained them from the consideration of the *unite hyperbola*. Also, he first introduced the notation **sh** and **ch**. Further investigation of the properties of the hyperbolic functions was carried out by Lambert.

It should be noted that the interest in the *hyperbolic functions* in mathematics increased sharply in the 19th century, when the Russian mathematician Nikolay Lobachevsky developed a new kind of geometry based on the *hyperbolic functions*; this kind of geometry is called *hyperbolic geometry* or *Lobachevsky's geometry*.

As it is known, *Lobachevsky's geometry* goes back in its origins to Euclid's *Elements* and is connected with the new formulation of the *Fifth Euclidean postulate* (the *postulate about parallel lines*).

We will not dwell on all the features of *Lobachevsky's geometry*. It is important to emphasize that, by investigating the trigonometric relations of his geometry, Lobachevsky used the *hyperbolic functions* introduced above (2.11), that is, *Lobachevsky's geometry proved to be the important confirmation of the fundamental nature of the hyperbolic functions for the geometric space surrounding us.*

2.6. Millennium Problems in Mathematics and Physics

In the recent years, the so-called *Millennium Problems* became a big passion of mathematicians and physicists. The outstanding mathematician David Hilbert signaled the beginning of this passion. In 1900, Hilbert presented the 23 Great Mathematical Problems at the *International Congress of Mathematicians in Paris.*

By explaining the purpose of the formulation of his "Mathematical Problems", David Hilbert writes thus [146]:

> *"For the close of a great epoch not only invites us to look back into the past but also directs our thoughts to the unknown future."*

As outlined in Ref. [147], *"Hilbert's 1900 address to the International Congress of Mathematicians in Paris was perhaps the most influential speech, given some day by a mathematician to mathematicians. In his speech, Hilbert outlined the 23 major mathematical problems, which to be studied in the coming century... Hilbert's address was more than a simple collection of problems. Hilbert outlined his philosophy of mathematics and proposed the problems, important for his philosophy."*

Modern mathematicians decided to continue the great tradition of David Hilbert. In May 2000, emulating Hilbert, the Clay Mathematics Institute of Cambridge announced (in Paris, for full effect) the *seven "Millennium Prize Problems"*, each with a bounty of $1 million [149].

Modern physicists have decided not to lag in comparison to mathematicians. They have formulated the *10 Physics Problems for the Next Millennium* [149].

2.7. A New Look at the Binet Formulas

2.7.1. Extended Fibonacci and Lucas numbers

In Vol. I, we introduced the notion of the *extended Fibonacci and Lucas numbers*. Let's recall these concepts. Fibonacci and Lucas numbers, F_n and L_n extended toward the negative values of the index n, are presented in Table 2.1.

We emphasize that the extended Fibonacci and Lucas numbers are the two infinite numerical sequences that change in limits from $-\infty$ to $+\infty$ because the index n takes the values from the set $n \in \{0, \pm 1, \pm 2, \pm 3, \dots\}$.

Table 2.1. Extended Fibonacci and Lucas numbers.

n	0	1	2	3	4	5	6	7	8	9	10
F_n	0	1	1	2	**3**	5	**8**	13	**21**	34	**55**
F_{-n}	0	1	−1	2	−**3**	5	−**8**	13	−**21**	34	−**55**
L_n	2	**1**	3	**4**	7	**11**	18	**29**	47	**76**	123
L_{-n}	2	−**1**	3	−**4**	7	−**11**	18	−**29**	47	−**76**	123

As it follows from Table 2.1, the elements of the extended sequences F_n and L_n have a number of remarkable mathematical properties. For example, for the *odd* indices $n = 2k + 1$, the elements of the sequences F_n and F_{-n} coincide, that is,

$$F_{2k+1} = F_{-2k-1}, \qquad (2.16)$$

but for the *even* indices $n = 2k$, they are opposite in sign, that is,

$$F_{2k} = -F_{-2k}. \qquad (2.17)$$

Note that the Fibonacci numbers F_n with the *even* indices $n = 2k$ are marked in bold in Table 2.1.

As for the Lucas numbers L_n, here we have opposite situation, that is, for the *even* indices $n = 2k$,

$$L_{2k} = L_{-2k}, \qquad (2.18)$$

but for the *odd* indices $n = 2k + 1$

$$L_{2k+1} = -L_{-2k-1}. \qquad (2.19)$$

Note that the Lucas numbers with the *odd* indices $n = 2k + 1$ are marked in bold in Table 2.1.

2.7.2. Deduction of the Binet formulas

To derive the *Binet formulas*, we use the remarkable identity, linking the neighboring powers of the golden proportion:

$$\Phi^n = \Phi^{n-1} + \Phi^{n-2}, \qquad (2.20)$$

where $n = 0, \pm 1, \pm 2, \pm 3, \ldots$.

We first write the expressions for the minus-one, zero and plus-one degrees of the *golden proportion* in the "explicit form":

$$\Phi^{-1} = \frac{-1 + \sqrt{5}}{2}, \quad \Phi^0 = 1 = \frac{2 + 0 \times \sqrt{5}}{2}, \quad \Phi^1 = \frac{1 + \sqrt{5}}{2}.$$

$$(2.21)$$

By using the initial values for the degrees of the *golden proportion*, given by (2.21), and applying the identity (2.20), we can represent the second, third and fourth degrees of the *golden proportion* in the "explicit form" as follows:

$$\Phi^2 = \Phi^1 + \Phi^0 = \frac{3 + \sqrt{5}}{2}, \quad \Phi^3 = \Phi^2 + \Phi^1 = \frac{4 + 2\sqrt{5}}{2},$$

$$\Phi^4 = \Phi^3 + \Phi^2 = \frac{7 + 3\sqrt{5}}{2}.$$

$$(2.22)$$

Can we see some regularity in the formulas (2.21) and (2.22)? First of all, we note that all the expressions (2.22) have one and the same mathematical structure:

$$\frac{A + B\sqrt{5}}{2}.$$

$$(2.23)$$

What are the numerical sequences A and B in these formulas? If we start with the expression for the zeroth power of the golden proportion Φ^0, it is easy to verify that the series of numbers A is a sequence of the numbers 2, 1, 3, 4, 7, 11, 18,... (Lucas numbers L_n) and the series of numbers B is the following numerical sequence: 0, 1, 1, 2, 3, 5, 8, ... (Fibonacci numbers F_n). From these examples, we can assume that in the general case, the formula, which allows expressing arbitrary (nth) degree of the *golden proportion* in terms of Lucas (L_n) and Fibonacci (F_n) numbers has the following form:

$$\Phi^n = \frac{L_n + F_n \sqrt{5}}{2}.$$

$$(2.24)$$

The formula (2.24) is easily proved by induction. Indeed, for $n = 1$, the formula (2.24) is valid because

$$\Phi^1 = \frac{L_1 + F_1\sqrt{5}}{2} = \frac{1 + \sqrt{5}}{2}. \tag{2.25}$$

The basis of the induction is proven.

If the formula (2.24) is valid for the given integers $(1, 2, \ldots, n - 2, n - 1, n)$, then this inductive hypothesis implies its validity for the integer $(n+1)$ because

$$\Phi^{n+1} = \frac{L_{n+1} + F_{n+1}\sqrt{5}}{2} = \frac{L_n + F_n\sqrt{5}}{2}$$
$$+ \frac{L_{n-1} + F_{n-1}\sqrt{5}}{2} = \Phi^n + \Phi^{n-1}, \tag{2.26}$$

which corresponds to the identity (2.20), connecting the degrees of the *golden proportion* Φ.

2.7.3. A new look on the Binet formulas for the Fibonacci and Lucas numbers

By using the formula (2.24), it is also very easy to express the extended Lucas numbers L_n and the extended Fibonacci numbers F_n through the *golden proportion*. To do this, it is enough to use the formula (2.24) and write down the expressions for the sum or difference of the nth degrees of the golden proportion $\Phi^n + \Phi^{-n}$ and $\Phi^n - \Phi^{-n}$ as follows:

$$\Phi^n + \Phi^{-n} = \frac{(L_n + L_{-n}) + (F_n + F_{-n})\sqrt{5}}{2}, \tag{2.27}$$

$$\Phi^n - \Phi^{-n} = \frac{(L_n - L_{-n}) + (F_n - F_{-n})\sqrt{5}}{2}. \tag{2.28}$$

Then, if we take into account the properties of the "extended Fibonacci and Lucas numbers", given by (2.16)–(2.19), then for the *even n = 2k,* we can rewrite the formulas (2.27) and (2.28) as follows:

$$\Phi^{2k} + \Phi^{-2k} = L_{2k}, \tag{2.29}$$

$$\Phi^{2k} - \Phi^{-2k} = F_{2k}\sqrt{5}. \tag{2.30}$$

Then, if we take into account the properties of the "extended Fibonacci and Lucas numbers", given by (2.16)–(2.19), then for the *even n = 2k,* we can rewrite the formulas (2.27) and (2.28) as follows:

$$\Phi^{2k+1} - \Phi^{-(2k+1)} = F_{2k+1}\sqrt{5}, \tag{2.31}$$

$$\Phi^{2k+1} + \Phi^{-(2k+1)} = L_{2k+1}. \tag{2.32}$$

The formulas (2.29)–(2.32) can be represented in the following compact form:

$$L_n = \begin{cases} \Phi^n + \Phi^{-n} & \text{for } n = 2k, \\ \Phi^n - \Phi^{-n} & \text{for } n = 2k+1, \end{cases} \tag{2.33}$$

$$F_n = \begin{cases} \dfrac{\Phi^n + \Phi^{-n}}{\sqrt{5}} & \text{for } n = 2k+1, \\[2ex] \dfrac{\Phi^n - \Phi^{-n}}{\sqrt{5}} & \text{for } n = 2k. \end{cases} \tag{2.34}$$

Analysis of the formulas (2.33) and (2.34) gives us the opportunity to get true aesthetic pleasure and once again be convinced in the power of human mind. Indeed, we know that the Fibonacci and Lucas numbers are always integers. On the other hand, any degree of the *golden proportion* is an irrational number. It follows from this that the integers L_n and F_n with the help of the formulas (2.33) and (2.34) can be very simply expressed through special irrational numbers (the *golden proportion* $\Phi = \frac{1+\sqrt{5}}{2}$ and $\sqrt{5}$).

2.8. Hyperbolic Fibonacci and Lucas Functions

2.8.1. A brief history

Usually in mathematics, the Binet formulas are represented in a form slightly different from (2.33) and (2.34). Traditional representations for the Binet formulas were considered in Vol. II. For the first time, the representation of these formulas in the new forms (2.33) and (2.34) was used by Alexey Stakhov in the book *Codes of the Golden Proportion* [19].

But it is precisely in this form, that the extended Fibonacci and Lucas numbers clearly show their "hyperbolic character" because the Binet formulas, represented in the forms (2.33) and (2.34), turn out to be similar in their mathematical structure to the formulas (2.11), which define the classical hyperbolic functions. It is this analogy that underlies the new class of hyperbolic functions called the *hyperbolic Fibonacci and Lucas functions*.

For the first time, a new class of hyperbolic functions, based on Binet's formulas, was introduced by the Ukrainian mathematicians Alexey Stakhov and Ivan Tkachenko in the late 1980s. The first article, describing this mathematical result, was published in 1988 as a preprint.

In 1993, according to the recommendation of Academician Mitropolsky, the article "Hyperbolic Fibonacci Trigonometry" [64] by Alexey Stakhov and Ivan Tkachenko was published in the Ukrainian academic journal, *Reports of the Academy of Sciences of Ukraine* [64] and thanks to this publication, a new class of hyperbolic functions entered into modern science and mathematics.

2.8.2. Symmetric hyperbolic Fibonacci and Lucas functions

The theory of hyperbolic Fibonacci and Lucas functions were further developed in the article by Alexey Stakhov and Boris Rozin, published in the international journal *Chaos, Solitons & Fractals* [75]. This article introduces the so-called *symmetric hyperbolic Fibonacci and Lucas functions* into modern mathematics.

Symmetric hyperbolic Fibonacci sine:

$$\mathrm{sFs}(x) = \frac{\Phi^x - \Phi^{-x}}{\sqrt{5}}, \qquad (2.35)$$

Symmetric hyperbolic Fibonacci cosine:

$$cFs(x) = \frac{\Phi^x + \Phi^{-x}}{\sqrt{5}}, \qquad (2.36)$$

Symmetric hyperbolic Lucas sine:

$$sLs(x) = \Phi^x - \Phi^{-x}, \qquad (2.37)$$

Symmetric hyperbolic Lucas cosine:

$$cLs(x) = \Phi^x + \Phi^{-x}, \qquad (2.38)$$

where x is a continuous variable, which takes its values in the range from $-\infty$ to $+\infty$.

By comparing the Binet formulas (2.33) and (2.34) to the formulas (2.35)–(2.38), it is easy to see that at the discrete points of the variable x, the symmetric hyperbolic Fibonacci and Lucas functions coincide with the Binet formulas (2.33) and (2.34):

$$F_n = \begin{cases} sFs(n) & \text{for } n = 2k, \\ cFs(n) & \text{for } n = 2k+1, \end{cases} \qquad (2.39)$$

$$L_n = \begin{cases} cLs(n) & \text{for } n = 2k, \\ sLs(n) & \text{for } n = 2k+1, \end{cases} \qquad (2.40)$$

where k takes the values from the set $k \in \{0, \pm 1, \pm 2, \pm 3, \ldots\}$.

It follows from the formula (2.39) that for all the *even* values of $n = 2k$, the values of the *hyperbolic Fibonacci sine* $sFs(n) = sFs(2k)$ coincide with the Fibonacci numbers for the *even* indices $F_n = F_{2k}$, and for all the *odd* values of $n = 2k + 1$, the values of the *hyperbolic Fibonacci cosine* $cFs(n) = cFs(2k + 1)$ coincide with the Fibonacci numbers for the *odd* indices $F_n = F_{2k+1}$. At the same time, it follows from the formula (2.40) that, for all the *even* values of $n = 2k$, the values of the *hyperbolic Lucas cosine* $cLs(n) = cLs(2k)$ coincide with the Lucas numbers for the *even* indices $L_n = L_{2k}$, and for all the *odd* values of $n = 2k + 1$, the values of the *hyperbolic Lucas sinus* $sLs(n) = sLs(2k + 1)$ coincide for the Lucas numbers with the *odd* indices $L_n = L_{2k+1}$. That is, the Fibonacci and Lucas numbers fit into the hyperbolic Fibonacci and Lucas functions at the "discrete" points of the continuous variable $x = 0, \pm 1, \pm 2, \pm 3, \ldots$.

Thus, according to (2.39), the Fibonacci numbers with the *even* indices ($n = 2k$) always correspond to the *symmetric Fibonacci sine*, and the Fibonacci numbers with the *odd* indices ($n = 2k + 1$) correspond to the *symmetric Fibonacci cosine* cFs(x), while according to (2.40), the Lucas numbers with the *even* indices always correspond to the *symmetric Lucas cosine* cLs(x), and the Lucas numbers with the *odd* indices correspond to the *symmetric Lucas sine* sLs(x).

Note also that the *symmetric hyperbolic Fibonacci and Lucas functions*, introduced above, are related to each other by the following simple relations:

$$\text{sFs}(x) = \frac{\text{sLs}(x)}{\sqrt{5}}, \quad \text{cFs}(x) = \frac{\text{cLs}(x)}{\sqrt{5}}. \tag{2.41}$$

By analogy with the classical hyperbolic functions (2.12), other symmetric hyperbolic Fibonacci and Lucas functions can be introduced, in particular, the *Fibonacci and Lucas tangent and cotangent*:

$$\text{tFs}(x) = \frac{\text{sFs}(x)}{\text{cFs}(x)}, \quad \text{ctFs}(x) = \frac{\text{cFs}(x)}{\text{sFs}(x)}, \tag{2.42}$$

$$\text{tLs}(x) = \frac{\text{sLs}(x)}{\text{cLs}(x)}, \quad \text{ctLs}(x) = \frac{\text{cLs}(x)}{\text{sLs}(x)}. \tag{2.43}$$

2.8.3. Parity property

The main advantage of the *symmetric hyperbolic Fibonacci and Lucas functions* (2.35)–(2.38), introduced in [75], in comparison to similar functions, introduced in [64], is the preservation of the important *parity property*, which is easily proved if we use the formulas (2.35)–(2.38). Really,

$$\text{sFs}(-x) = \frac{\Phi^{-x} - \Phi^x}{\sqrt{5}} = -\frac{\Phi^x - \Phi^{-x}}{\sqrt{5}} = -\text{sFs}(x), \tag{2.44}$$

that is, the *symmetric hyperbolic Fibonacci sine* is the *odd* function.

Similarly, it is easy to prove that the *symmetric hyperbolic Fibonacci cosine* is the *even* function, that is,

$$cFs(-x) = \frac{\Phi^{-x} + \Phi^x}{\sqrt{5}} = \frac{\Phi^x + \Phi^{-x}}{\sqrt{5}} = cFs(x). \qquad (2.45)$$

In the same way, one can prove that

$$sL(-x) = -sL(x), \quad cL(-x) = cL(x). \qquad (2.46)$$

2.8.4. Graphs of the symmetric hyperbolic Fibonacci and Lucas functions

The graphs of the functions (2.35)–(2.38), shown in Figs. 2.9 and 2.10, according to the *parity properties* (2.44)–(2.46), have a symmetrical character and are similar to the graphs of the *classical hyperbolic functions* (Fig. 2.9).

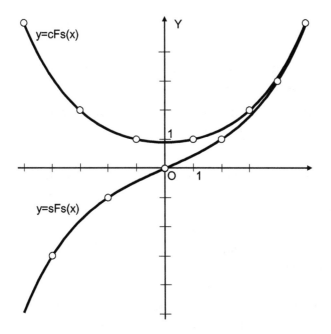

Fig. 2.9. Symmetric hyperbolic Fibonacci functions.

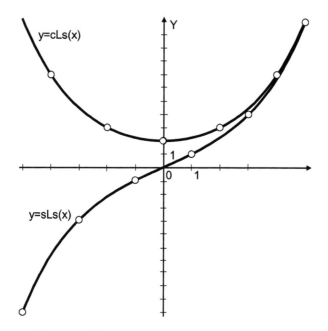

Fig. 2.10. Symmetric hyperbolic Lucas functions.

2.9. Recurrent Properties of the Hyperbolic Fibonacci and Lucas Functions

2.9.1. Analogy to the recurrent relations for Fibonacci and Lucas numbers

A detailed study of the mathematical properties of a new class of hyperbolic functions was carried out in Ref. [75]. It has been shown that the symmetric hyperbolic Fibonacci and Lucas functions, on the one hand, have *recurrent properties* similar to the properties of Fibonacci and Lucas numbers, on the other hand, they have *hyperbolic properties* similar to the classical hyperbolic functions (2.11).

Let's now prove some of the simplest recurrent properties of the symmetric hyperbolic Fibonacci and Lucas functions. For example, it is easy to prove that the Fibonacci recurrent relation $F_{n+2} = F_{n+1} + F_n$ corresponds to the following identities for the symmetric hyperbolic Fibonacci functions:

$$\mathrm{sFs}(x+2) = \mathrm{cFs}(x+1) + \mathrm{sFs}(x),$$
$$\mathrm{cFs}(x+2) = \mathrm{sFs}(x+1) + \mathrm{cFs}(x). \qquad (2.47)$$

Really,

$$\mathrm{cFs}(x+1) + \mathrm{sFs}(x)$$

$$= \frac{\Phi^{x+1} + \Phi^{-(x+1)}}{\sqrt{5}} + \frac{\Phi^x - \Phi^{-x}}{\sqrt{5}} = \frac{\Phi^x\,(\Phi+1) - \Phi^{-x}\,(1-\Phi)}{\sqrt{5}}$$

$$= \frac{\Phi^x \times \Phi^2 - \Phi^{-x} \times \Phi^{-2}}{\sqrt{5}} = \frac{\Phi^{x+2} - \Phi^{-(x+2)}}{\sqrt{5}} = \mathrm{sFs}\,(x+2),$$

$$\mathrm{sFs}(x+1) + \mathrm{cFs}(x)$$

$$= \frac{\Phi^{x+1} - \Phi^{-(x+1)}}{\sqrt{5}} + \frac{\Phi^x + \Phi^{-x}}{\sqrt{5}} = \frac{\Phi^x\,(\Phi+1) + \Phi^{-x}\,(1-\Phi)}{\sqrt{5}}$$

$$= \frac{\Phi^x \times \Phi^2 + \Phi^{-x} \times \Phi^{-2}}{\sqrt{5}} = \frac{\Phi^{x+2} + \Phi^{-(x+2)}}{\sqrt{5}} = \mathrm{cFs}(x+2).$$

Similarly, we can prove that the recurrent relation for the Lucas numbers $L_{n+2} = L_{n+1} + L_n$ corresponds to the following identities for the symmetric hyperbolic Lucas λ-functions:

$$\mathrm{sLs}(x+2) = \mathrm{cLs}(x+1) + \mathrm{sLs}(x),$$
$$\mathrm{cLs}(x+2) = \mathrm{sLs}(x+1) + \mathrm{cLs}(x). \qquad (2.48)$$

2.9.2. Generalization of the Cassini formula

As mentioned in Vol. II, the Cassini formula, $F_n^2 - F_{n+1}F_{n-1} = (-1)^{n+1}$, is one of the most surprising properties, connecting the three adjacent Fibonacci numbers.

Let's prove that this formula corresponds to the two identities for the symmetric hyperbolic Fibonacci functions:

$$[\mathrm{sFs}(x)]^2 - \mathrm{cFs}(x+1)cF(x-1) = -1,$$
$$[\mathrm{cFs}(x)]^2 - \mathrm{sFs}(x+1)sF(x-1) = 1. \qquad (2.49)$$

Really,

$$[\mathrm{sFs}(x)]^2 - \mathrm{cFs}(x+1)\,\mathrm{cFs}(x-1)$$

$$= \left(\frac{\Phi^x - \Phi^{-x}}{\sqrt{5}}\right)^2 - \frac{\Phi^{x+1} + \Phi^{-(x+1)}}{\sqrt{5}} \times \frac{\Phi^{x-1} + \Phi^{-(x-1)}}{\sqrt{5}}$$

$$= \frac{\Phi^{2x} - 2 + \Phi^{-2x} - \left(\Phi^{2x} + \Phi^2 + \Phi^{-2} + \Phi^{-2x}\right)}{5} = -1,$$

$$[\mathrm{cFs}(x)]^2 - \mathrm{sFs}(x+1)\,\mathrm{sFs}(x-1)$$

$$= \left(\frac{\Phi^x + \Phi^{-x}}{\sqrt{5}}\right)^2 - \frac{\Phi^{x+1} - \Phi^{-(x+1)}}{\sqrt{5}} \times \frac{\Phi^{x-1} - \Phi^{-(x-1)}}{\sqrt{5}}$$

$$= \frac{\Phi^{2x} + 2 + \Phi^{-2x} - \left(\Phi^{2x} - \Phi^2 - \Phi^{-2} + \Phi^{-2x}\right)}{5} = 1.$$

Note that in the proof of these identities, we used the Binet formula for Lucas numbers:

$$\Phi^2 + \Phi^{-2} = L_3 = 3.$$

2.9.3. Table of the recurrent properties of the symmetric hyperbolic Fibonacci and Lucas functions

In the same way, the new identities for the *symmetric hyperbolic Fibonacci and Lucas functions* [75] can be derived from the well-known identities for the Fibonacci and Lucas numbers. Some of them are given in Table 2.2.

It follows from Table 2.2 that for each recurrent relation for the Fibonacci and Lucas numbers, there correspond the two identities for the symmetric hyperbolic Fibonacci and Lucas functions. This fact is explained very simply. The choice of one of the two identities for the hyperbolic Fibonacci and Lucas functions depends on the *parity* of the index n in the Fibonacci and Lucas numbers.

Let's consider, for example, the simplest identities corresponding to the recurrent Fibonacci relation $F_{n+2} = F_{n+1} + F_n$. It follows from

Table 2.2. Recurrent properties of the symmetric hyperbolic Fibonacci and Lucas functions.

Identities for Fibonacci and Lucas numbers	Identities for the symmetric hyperbolic Fibonacci and Lucas functions
$F_{n+2} = F_{n+1} + F_n$	$sFs(x+2) = cFs(x+1) + sFs(x),\ cFs(x+2) = sFs(x+1) + cFs(x)$
$L_{n+2} = L_{n+1} + L_n$	$sLs(x+2) = cLs(x+1) + sLs(x),\ cLs(x+2) = sLs(x+1) + cLs(x)$
$F_n = (-1)^{n+1} F_{-n}$	$sFs(x) = -sFs(-x),\ cFs(x) = cFs(-x)$
$L_n = (-1)^n L_{-n}$	$sLs(x) = -sLs(-x),\ cLs(x) = cLs(-x)$
$F_{n+3} + F_n = 2F_{n+2}$	$sFs(x+3) + cFs(x) = 2cFs(x+2),\ cFs(x+3) + sFs(x) = 2sFs(x+2)$
$F_{n+3} - F_n = 2F_{n+1}$	$sFs(x+3) - cFs(x) = 2sFs(x+1),\ cFs(x+3) - sFs(x) = 2cFs(x+1)$
$F_{n+6} - F_n = 4F_{n+3}$	$sFs(x+6) - cFs(x) = 4cFs(x+3),\ cFs(x+6) - sFs(x) = 4cFs(x+3)$
$F_n^2 - F_{n+1}F_{n-1} = (-1)^{n+1}$	$[sFs(x)]^2 - cFs(x+1)cFs(x-1) = -1,\ [cFs(x)]^2 - sFs(x+1)sFs(x-1) = 1$
$F_{2n+1} = F_{n+1}^2 + F_n^2$	$cFs(2x+1) = [cFs(x+1)]^2 + [sFs(x)]^2,\ sFs(2x+1) = [sFs(x+1)]^2 + [cFs(x)]^2$
$L_n^2 - 2(-1)^n = L_{2n}$	$[sLs(x)]^2 + 2 = cLs(2x),\ [cLs(x)]^2 - 2 = cLs(2x)$
$L_n + L_{n+3} = 2L_{n+2}$	$sLs(x) + cLs(x+3) = 2sLs(x+2),\ cLs(x) + sLs(x+3) = 2cLs(x+2)$
$L_{n+1}L_{n-1} - L_n^2 = -5(-1)^n$	$sLs(x+1)sLs(x-1) - [cLs(x)]^2 = -5,\ cLs(x+1)cLs(x-1) - [sLs(x)]^2 = 5$
$F_{n+3} - 2F_n = L_n$	$sFs(x+3) - 2cFs(x) = sLs(x),\ cFs(x+3) - 2sFs(x) = cLs(x)$
$L_{n-1} + L_{n+1} = 5F$	$sLs(x-1) + cLs(x+1) = 5sFs(x),\ cLs(x-1) + sLs(x+1) = 5cFs(x)$
$L_n + 5F_n = 2L_{n+1}$	$sLs(x) + 5cFs(x) = 2cLs(x+1),\ cLs(x) + 5sFs(x) = 2sLs(x+1)$
$L_{n+1}^2 + L_n^2 = 5F_{2n+1}$	$[sLs(x+1)]^2 + [cLs(x)]^2 = 5cFs(2x+1),\ [cLs(x+1)]^2 + [sLs(x)]^2 = 5sFs(2x+1)$

Table 2.2 that the two identities

(1) $\text{sFs}(x+2) = \text{cFs}(x+1) + \text{sFs}(x)$,
(2) $\text{cFs}(x+2) = \text{sFs}(x+1) + \text{cFs}(x)$

correspond to the recurrent relation $F_{n+2} = F_{n+1} + F_n$. If $n = 2k$, we should use the first identity $\text{sFs}(x+2) = \text{cFs}(x+1) + \text{sFs}(x)$; for the case $n = 2k + 1$ we should use the second identity $\text{cFs}(x+2) = \text{sFs}(x+1) + \text{cFs}(x)$.

2.9.4. The theory of the symmetric hyperbolic Fibonacci and Lucas functions as the "golden" paradigm of the "Fibonacci numbers theory"

As shown above, two "continuous" identities for the hyperbolic Fibonacci and Lucas functions correspond to one recurrent relation for the Fibonacci and Lucas numbers. Because the Fibonacci and Lucas numbers according to (2.39), (2.40) are the "discrete" case of the symmetric hyperbolic Fibonacci and Lucas functions, with which they coincide for the "discrete" values of the continuous variable $x = 0, \pm1, \pm2, \pm3, \ldots$, then this means that with the introduction of the *symmetric hyperbolic Fibonacci and Lucas functions*, the *classical theory of Fibonacci numbers* [8, 9, 11] as if "degenerates" because it becomes a particular ("discrete") case of the more general ("continuous") theory of the *symmetric hyperbolic Fibonacci and Lucas functions*.

Such a look at the theory of Fibonacci and Lucas numbers reflects the new paradigm (the *"golden" paradigm*) in the development of modern theory of Fibonacci and Lucas numbers; such the "golden" paradigm is extending the subject of the "theory of Fibonacci and Lucas numbers", especially in its applications to theoretical natural sciences.

2.10. Hyperbolic Properties of the Symmetric Hyperbolic Fibonacci and Lucas Functions

2.10.1. Parity property

The symmetric hyperbolic Fibonacci and Lucas functions, introduced above, preserve all the well-known properties of the classical hyperbolic functions (2.11). The *parity property* (2.44)–(2.46), considered above, is the most important of them. Indeed, the *Fibonacci and Lucas hyperbolic sines* are the *odd* functions, while the *Fibonacci and Lucas hyperbolic cosines* are the *even* functions:

$$\begin{aligned} \mathrm{sFs}(-x) &= -\mathrm{sFs}(x), & \mathrm{cFs}(-x) &= \mathrm{cFs}(x), \\ \mathrm{sLs}(-x) &= -\mathrm{sLs}(x), & \mathrm{cLs}(-x) &= \mathrm{cLs}(x). \end{aligned} \tag{2.50}$$

2.10.2. Fundamental identities for the symmetric hyperbolic Fibonacci and Lucas functions

But there are deeper mathematical connections between the *classical hyperbolic functions* and the *symmetric hyperbolic Fibonacci and Lucas functions*. For example, one of the most important properties of the *classical hyperbolic functions* is the following identity:

$$[\mathrm{ch}(x)]^2 - [\mathrm{sh}(x)]^2 = 1. \tag{2.51}$$

It is easy to prove that in terms of the *symmetric hyperbolic Fibonacci and Lucas functions*, this identity can be represented in the form of two identities (for the symmetric hyperbolic Fibonacci and Lucas functions):

$$[\mathrm{cFs}(x)]^2 - [\mathrm{sFs}(x)]^2 = \frac{4}{5}, \tag{2.52}$$

$$[\mathrm{cLs}(x)]^2 - [\mathrm{sLs}(x)]^2 = 4. \tag{2.53}$$

Let's prove the identity (2.52):

$$\begin{aligned} [\mathrm{cFs}(x)]^2 - [\mathrm{sFs}(x)]^2 &= \left(\frac{\Phi^x + \Phi^{-x}}{\sqrt{5}}\right)^2 - \left(\frac{\Phi^x - \Phi^{-x}}{\sqrt{5}}\right)^2 \\ &= \frac{\Phi^{2x} + 2 + \Phi^{-2x} - \Phi^{2x} + 2 - \Phi^{-2x}}{5} = \frac{4}{5}. \end{aligned}$$

The identity (2.53) is proved similarly.

2.10.3. The table of the identities for the symmetric hyperbolic Fibonacci and Lucas functions

As mentioned above, for each identity for the *classical hyperbolic functions*, there is an analogue in the form of the corresponding identity for the *symmetric hyperbolic Fibonacci and Lucas functions.* Table 2.3 shows some formulas for the *classical hyperbolic functions* and the corresponding formulas for the *symmetric hyperbolic Fibonacci functions.* Similar formulas for the *symmetric hyperbolic Lucas functions* can be obtained from Table 2.3 if we use the relations (2.41), connecting the *symmetric hyperbolic Fibonacci and Lucas functions* among themselves.

2.11. Formulas for Differentiation and Integration

The following formulas for differentiation of the classical hyperbolic functions are well known:

$$[\mathrm{ch}(x)]^{(n)} = \begin{cases} \mathrm{sh}(x) & \text{for } n = 2k + 1, \\ \mathrm{ch}(x) & \text{for } n = 2k, \end{cases}$$

$$[\mathrm{sh}(x)]^{(n)} = \begin{cases} \mathrm{ch}(x) & \text{for } n = 2k + 1, \\ \mathrm{sh}(x) & \text{for } n = 2k. \end{cases}$$

It is easy to prove the following formulas for differentiation of the symmetric hyperbolic Fibonacci functions [75]:

$$[\mathrm{cFs}(x)]^{(n)} = \begin{cases} (\ln \Phi)^n \, \mathrm{sFs}(x), & \text{for } n = 2k + 1, \\ (\ln \Phi)^n \, \mathrm{cFs}(x), & \text{for } n = 2k, \end{cases}$$

$$[\mathrm{sFs}(x)]^{(n)} = \begin{cases} (\ln \Phi)^n \, \mathrm{cFs}(x) & \text{for } n = 2k + 1, \\ (\ln \Phi)^n \, \mathrm{sFs}(x) & \text{for } n = 2k. \end{cases}$$

The formulas for integration of the *classical hyperbolic functions* have the following forms:

$$\underbrace{\int\!\!\int\!\!\int}_{n} \mathrm{ch}(x)dx = \begin{cases} \mathrm{sh}(x), & \text{for } n = 2k + 1, \\ \mathrm{ch}(x), & \text{for } n = 2k, \end{cases}$$

$$\underbrace{\int\!\!\int\!\!\int}_{n} \mathrm{sh}(x)dx = \begin{cases} \mathrm{ch}(x), & \text{for } n = 2k + 1, \\ \mathrm{sh}(x), & \text{for } n = 2k. \end{cases}$$

Table 2.3. "Hyperbolic" properties of the symmetric hyperbolic Fibonacci functions.

Formulas for the classical hyperbolic functions	Formulas for the symmetrical hyperbolic Fibonacci functions
$\mathrm{sh}(x) = \frac{e^x - e^{-x}}{2},\ \mathrm{ch}(x) = \frac{e^x + e^{-x}}{2}$	$\mathrm{sF}_\lambda(x) = \frac{\Phi_\lambda^x - \Phi_\lambda^{-x}}{\sqrt{4+\lambda^2}},\ c F_\lambda(x) = \frac{\Phi_\lambda^x + \Phi_\lambda^{-x}}{\sqrt{4+\lambda^2}}$
$\mathrm{ch}^2(x) - \mathrm{sh}^2(x) = 1$	$[\mathrm{cFs}(x)]^2 - [\mathrm{sFs}(x)]^2 = \frac{4}{5}$
$\mathrm{sh}(x+y) = \mathrm{sh}(x)\,\mathrm{ch}(x) + \mathrm{ch}(x)\,\mathrm{sh}(x)$	$\frac{2}{\sqrt{5}}\mathrm{sFs}(x+y) = \mathrm{sFs}(x)\,\mathrm{cFs}(x) + \mathrm{cFs}(x)\,\mathrm{sFs}(x)$
$\mathrm{sh}(x-y) = \mathrm{sh}(x)\,\mathrm{ch}(x) - \mathrm{ch}(x)\,\mathrm{sh}(x)$	$\frac{2}{\sqrt{5}}\mathrm{sFs}(x-y) = \mathrm{sFs}(x)\,\mathrm{cFs}(x) - \mathrm{cFs}(x)\,\mathrm{sFs}(x)$
$\mathrm{ch}(x+y) = \mathrm{ch}(x)\,\mathrm{ch}(x) + \mathrm{sh}(x)\,\mathrm{sh}(x)$	$\frac{2}{\sqrt{5}}\mathrm{cFs}(x+y) = \mathrm{cFs}(x)\,\mathrm{cFs}(x) + \mathrm{sFs}(x)\,\mathrm{sFs}(x)$
$\mathrm{ch}(x-y) = \mathrm{ch}(x)\,\mathrm{ch}(x) - \mathrm{sh}(x)\,\mathrm{sh}(x)$	$\frac{2}{\sqrt{5}}\mathrm{cFs}(x-y) = \mathrm{cFs}(x)\,\mathrm{cFs}(x) - \mathrm{sFs}(x)\,\mathrm{sFs}(x)$
$\mathrm{ch}(2x) = 2\,\mathrm{sh}(x)\,\mathrm{ch}(x)$	$\frac{1}{\sqrt{5}}\mathrm{cFs}(2x) = \mathrm{sFs}(x)\,\mathrm{cFs}(x)$
$[\mathrm{ch}(x) \pm \mathrm{sh}(x)]^n = \mathrm{ch}(nx) \pm \mathrm{sh}(nx)$	$[\mathrm{cFs}(x) \pm \mathrm{sFs}(x)]^n = \left(\frac{2}{\sqrt{5}}\right)^{n-1}[\mathrm{cFs}(nx) \pm \mathrm{sFs}(nx)]$

It is easy to derive the following formulas for integration of the symmetric hyperbolic Fibonacci and Lucas functions [75]:

$$\iiint_n \mathrm{cFs}\,(x)dx = \begin{cases} (\ln \Phi)^{-n}\,\mathrm{sFs}\,(x) & \text{for } n = 2k+1, \\ (\ln \Phi)^{-n}\,\mathrm{cFs}\,(x) & \text{for } n = 2k, \end{cases}$$

$$\iiint_n \mathrm{cLs}\,(x)dx = \begin{cases} (\ln \Phi)^{-n}\,\mathrm{sLs}\,(x) & \text{for } n = 2k+1, \\ (\ln \Phi)^{-n}\,\mathrm{cLs}\,(x) & \text{for } n = 2k, \end{cases}$$

$$\iiint_n \mathrm{sFs}\,(x)dx = \begin{cases} (\ln \Phi)^{-n}\,\mathrm{cFs}\,(x) & \text{for } n = 2k+1, \\ (\ln \Phi)^{-n}\,\mathrm{sFs}\,(x) & \text{for } n = 2k, \end{cases}$$

$$\iiint_n \mathrm{sLs}\,(x)dx = \begin{cases} (\ln \Phi)^{-n}\,\mathrm{cLs}\,(x) & \text{for } n = 2k+1, \\ (\ln \Phi)^{-n}\,\mathrm{sLs}\,(x) & \text{for } n = 2k. \end{cases}$$

2.11.1. Aesthetics of the symmetric hyperbolic Fibonacci and Lucas functions

Thus, the *symmetric hyperbolic functions*, introduced above, completely preserve the properties of the *classical hyperbolic functions* (Table 2.3), but at the same time, they have new (*"recurrent"*) properties similar to the properties of the Fibonacci and Lucas numbers (Table 2.2). At the same time, unlike the *classical hyperbolic functions*, the new hyperbolic functions have a "discrete analog" in the form of Fibonacci and Lucas numbers; according to (2.39) and (2.40), these new hyperbolic functions coincide with Fibonacci and Lucas numbers when the continuous variable x takes the "discrete" values: $x = 0, \pm 1, \pm 2, \pm 3, \ldots$. We also note that the relations (2.47)–(2.50), (2.52), and (2.53), as well as the other identities, given in Tables 2.2 and 2.3, are aesthetically perfect and satisfy the *Dirac principle of "mathematical beauty"*, which emphasizes the fundamental nature of the *symmetric hyperbolic Fibonacci and Lucas functions*, introduced above, and a conception of the *symmetric hyperbolic Fibonacci and Lucas functions* as the *"golden" paradigm* of the *symmetric hyperbolic Fibonacci and Lucas functions*.

Chapter 3

Applications of the Symmetric Hyperbolic Fibonacci and Lucas Functions

3.1. New Geometric Theory of Phyllotaxis ("Bodnar Geometry")

3.1.1. The riddle of phyllotaxis

In Vol. II, we described the botanical *phenomenon of phyllotaxis*, which consists in a regular spiral-symmetric arrangement of botanical objects. On the surfaces of such objects, its bio-organs (shoots of plants and trees, seeds on sunflower discs, scales of coniferous cones and pineapples, etc.) are arranged in the form of ordered patterns, which are formed by intersection of the left and right curved spiral lines, *parastichies*.

To characterize the phyllotaxis of such botanical objects, the number of left and right spirals observed on the surface of the phyllotaxis objects is usually indicated; at the same time, the patterns of phyllotaxis of such structures are described by the relations of the *neighboring Fibonacci numbers*:

$$\frac{F_{n+1}}{F_n}: \frac{2}{1}, \frac{3}{2}, \frac{5}{3}, \frac{8}{5}, \frac{13}{8}, \frac{21}{13}, \ldots \qquad (3.1)$$

These relations are called the *order of symmetry* of a phyllotaxis object and are characteristic of each phyllotaxis object [28]. For example, for the disc's of sunflowers (Fig. 3.1), the orders of symmetry can reach the values $\frac{89}{55}$, $\frac{144}{89}$ and even $\frac{233}{144}$.

Fig. 3.1. Sunflower disk.

By observing the phyllotaxis objects in the completed state and enjoying the orderly pattern on their surface, we always ask the question: How are the *Fibonacci lattices* formed on its surface during the growth process? This problem is one of the most intriguing *mysteries of phyllotaxis*. Its essence is that in most species of bioforms in the process of growth, there is a *change of the numerical characteristics of symmetry*.

It is known, for example, that sunflower disks, located at the different levels of the same stem, have different symmetries: *the older the disk, the higher the order of its symmetry* (Fig. 3.1). This means that in the process of growth, a regular change (increase) of the order of symmetry occurs and this change is carried out according to the regularity:

$$\frac{2}{1} \to \frac{3}{2} \to \frac{5}{3} \to \frac{8}{5} \to \frac{13}{8} \to \frac{21}{13} \to \cdots . \qquad (3.2)$$

Changing the symmetry orders of phyllotaxis objects in accordance with (3.2) is called *dynamic symmetry* [28]. All the above data constitute the essence of the well-known *mystery of phyllotaxis*. A number of scientists, who have studied this problem, suggest that the phenomenon of phyllotaxis has a fundamental interdisciplinary importance. According to Vernadsky, the *problem of biological symmetry is a key problem in biology*.

So, the phenomenon of *dynamic symmetry* (3.2) reveals its special role in the geometric problem of phyllotaxis. This suggests that the numerical regularity (3.2) hides certain geometric laws, which, perhaps, are the essence of the secret of the growth mechanism of phyllotaxis and their disclosure would be of great importance not only for botany but also for biology.

By exploring the phenomenon of *dynamic symmetry*, the Ukrainian researcher Oleg Bodnar made a scientific discovery [28]. He proved that the *geometry of phyllotaxis is non-Euclidean*; this geometry is a special type of hyperbolic geometry based on the "golden" hyperbolic functions.

3.1.2. Structural and numerical analysis of the phyllotaxis lattices

Consider a cedar cone as a characteristic natural phyllotaxis object (Fig. 3.2(a)). On the surface of the cone, each element of scales is

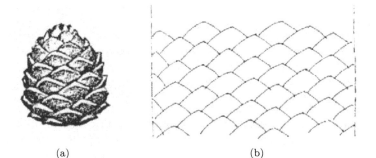

(a)   (b)

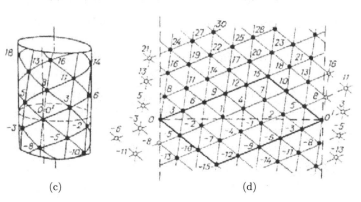

(c)   (d)

Fig. 3.2. Analysis of the structural and numerical properties of the phyllotaxis lattice.

blocked in three directions; as a result, a drawing is created from three types of spirals, the number of which is equal to the *Fibonacci numbers*: 3, 5, 8.

To simplify the geometric model of the phyllotaxis object in Fig. 3.2(a), we will represent it in the form of a cylinder (Fig. 3.2(c)). A mosaic ornament, created by the arrangement of the scales, will be represented in the form of a lattice, built on the centers of the scales. By cutting the cylinder along a vertical straight line (generator) and turning it onto a plane, we obtain a fragment of the phyllotaxis lattice, bounded by the two parallel straight lines, the *traces* of the cut line (Fig. 3.2(d)). Three types of spirals of the cylindrical surface of the lattice correspond to three groups of parallel straight lines: the three straight lines 0–21, 1–16, 2–8 have a right slope, the five straight lines 3–8, 1–16, 4–19, 7–27, 0–30 have a left slope, and the eight steep straight lines 0–24, 3–27, 6–30, 1–25, 4–25, 7–28, 2–18, 5–21 have a right slope.

We use the following method of numbering the nodal points of the lattice. We take the straight line, passing through the points O and O' as the *axis of abscises*, and the trace of the vertical section, passing through the point O, as the *axis of ordinates*. We now take for the unit the length the ordinate of the point 1, then the number of any point of the lattice will be equal to its ordinate.

A numbered flat lattice has a number of characteristic properties. An arbitrary pair of vertices defines a certain direction in the lattice system and, ultimately, a set of parallel directions, which are characterized by the constancy of the difference d between the numbers of adjacent vertices. The value of d becomes an indicator of the number of spirals that correspond to the considered direction of the lattice on the surface of the cylinder. So, for directions, given by the sides of the elementary triangle of the phyllotaxis lattice, the values of d are equal to 3, 5, 8 which correspond to the quantitative composition of the three groups of spirals of the cylindrical surface. It is noteworthy that the numbers of the points, directly adjacent to the point O, are also equal to the numbers 3, 5, 8. Thus, the order of symmetry is reflected in the lattice numbering system.

Finally, we pay attention to the fact that the segment OO' can be considered as a diagonal of the parallelogram, built on the basis of directions, for which d is equal to 5 and 3. The dimensions of the sides of this parallelogram, expressed in the number of intervals between the lattice nodes, are equal to 3 and 5, respectively. Thus, this parallelogram allows us to estimate the lattice symmetry without resorting to digital numbering. Let's call this parallelogram a *coordinate parallelogram*. Lattices with different symmetries correspond to coordinate parallelograms of various sizes.

3.1.3. Key principle of dynamic symmetry

We now proceed to the direct analysis of the phenomenon of *dynamic symmetry*. The idea of the analysis reduces to a comparison of the series of lattices (cylinder's sweeps) with different symmetries.

The lattices, presented in Fig. 3.3, are considered as successive stages of transformation of one and the same phyllotaxis object in the process of its growth. The question is how the lattices are transformed, i.e., what geometric motion can ensure the successive passage by the lattices of all the illustrated stages.

Figure 3.3 illustrates the example of the *Fibonacci phyllotaxis*, in which the natural sequence of symmetry changes is observed: $1 : 2 : 1 \rightarrow 2 : 3 : 1 \rightarrow 2 : 5 : 3 \rightarrow 5 : 8 : 3 \rightarrow 5 : 13 : 8$.

Let us analyze the stages I and II in Fig. 3.3. To do this, we use the concept of *compression* (*stretching*) of the plane, which we introduced above, when we analyzed the important concept of hyperbolic geometry: the *hyperbolic rotation*. At this stage, the lattice can be transformed by *compressing* the plane along the direction O–3 to the position, when the segment O–3 will reach the *lattice edge*.

At the same time, we must extend the plane in the direction of 1–2, perpendicular to the direction of the *compression*. During the transition from the stage II to the stage III, the *compression* should be along the direction O–5, and the *stretching* along the direction 2–3. The next transition is accompanied by a similar plane deformation along the directions O–8 (*compression*) and 3–5 (*stretching*).

Let's now consider the parallelogram $O1O_1'\bar{1}$ in Fig. 3.3(I). It is clear that in the process of further transformation, this parallelogram is sequentially transformed into parallelograms $O1O_2'\bar{1}$, $O1O_3'\bar{1}$,

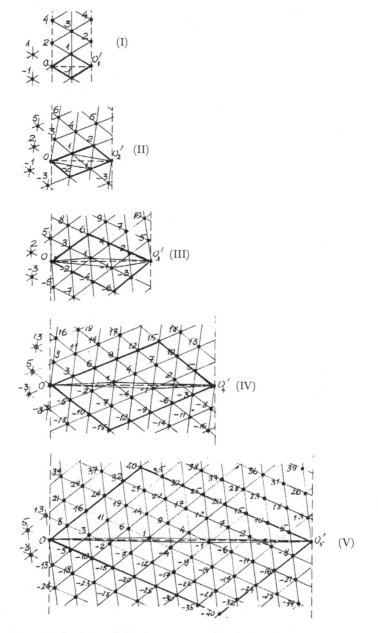

Fig. 3.3. Analysis of the dynamic symmetry of a phyllotaxis object.

$O1O_4'\bar{1}$, and $O1O_5'\bar{1}$ at the subsequent stages of the transformation of phyllotaxis lattices.

We will not delve deep into the detailed analysis of the scientific discovery of Oleg Bodnar by shifting the attention of the readers to Bodnar's books [28, 42]. We will only formulate the *key principle of the phenomenon of dynamic symmetry*. Bodnar argues that such a principle is a *hyperbolic rotation*. In accordance with this principle, the growth of the phyllotaxis object, which is accompanied by the change of the *dynamic symmetry*, can be modeled as a sequential transition from the object with the *lower symmetry order* to the object with the *higher symmetry order* (Fig. 3.3) which is carried out by means of the *hyperbolic rotation*. With this reasoning, Bodnar proved the *hyperbolic nature of the botanical phenomenon of phyllotaxis!*

3.1.4. The "golden" hyperbolic functions

Thus, the analysis of the mechanism for changing of the *dynamic symmetry* in the process of growth of the *phyllotaxis object* (Fig. 3.3) led Bodnar to the conclusion that the *geometry of phyllotaxis is hyperbolic*. However, it differs from hyperbolic geometry, based on classical hyperbolic functions, in the fact that it is implemented on the basis of another type of hyperbolic functions introduced by Oleg Bodnar [28, 42]. He called these functions the *"golden" hyperbolic functions*:

The "golden" hyperbolic sine:

$$\text{Gsh } n = \frac{\Phi^n - \Phi^{-n}}{2}, \tag{3.3}$$

The "golden" hyperbolic cosine:

$$\text{Gch } n = \frac{\Phi^n + \Phi^{-n}}{2}. \tag{3.4}$$

The coordinate transformation, called the *hyperbolic rotation*, in terms of the *"golden" hyperbolic functions* (3.3) and (3.4) is described by the following formulas:

$$x' = X \operatorname{Gch} n + Y \operatorname{Gsh} n, \quad y' = X \operatorname{Gsh} n + Y \operatorname{Gch} n.$$

Next, Bodnar establishes a fundamental connection between the *"golden" hyperbolic functions* (3.3) and (3.4), introduced by him, and the *Fibonacci numbers*:

$$F_{2k-1} = \frac{2}{\sqrt{5}} \operatorname{Gch}(2k - 1), \tag{3.5}$$

$$F_{2k} = \frac{2}{\sqrt{5}} \operatorname{Gsh} 2k. \tag{3.6}$$

By using the relations (3.5) and (3.6), Bodnar gives a surprisingly simple explanation of the *riddle of phyllotaxis*: Why do the *Fibonacci numbers* with persistent constancy appear on the surface of phyllotaxis objects? The main reason is the fact that wildlife geometry is *non-Euclidean geometry*; at the same time, it differs significantly from the Lobachevsky geometry and four-dimensional Minkowski geometry, based on the classical hyperbolic functions, primarily because the basic relations of this geometry are described by using the *golden hyperbolic functions* (3.3) and (3.4), which are connected to the *Fibonacci numbers* by simple relations (3.5) and (3.6).

It is with the use of the formulas (3.5) and (3.6) that Bodnar builds the original geometric theory of phyllotaxis related to the *"golden section"* and the *Fibonacci numbers*.

3.1.5. Connection of the "golden" hyperbolic functions with the symmetric hyperbolic Fibonacci functions

Earlier, we introduced the so-called *symmetric hyperbolic Fibonacci functions*, which are defined by the formulas (2.35) and (2.36) (see Chapter 2).

By comparing the formulas (2.35) and (2.36), which define the *symmetric hyperbolic Fibonacci functions*, with the formulas (3.3) and (3.4), which define the *"golden hyperbolic functions"*, we can establish the following relationship between the above groups of the formulas:

$$\text{Gsh}(x) = \frac{\sqrt{5}}{2}\,\text{sFs}(x), \qquad (3.7)$$

$$\text{Gch}(x) = \frac{\sqrt{5}}{2}\,\text{cFs}(x). \qquad (3.8)$$

Analysis of the formulas (3.7) and (3.8) allows us to conclude that the *"golden" sines and cosines*, introduced in Ref. [28], and the *symmetric hyperbolic Fibonacci functions*, introduced in Ref. [75], are essentially the same functions, which differ only with constant coefficients. At the same time, when Bodnar studies the *geometry of phyllotaxis* [28], he multiplies his *"golden" hyperbolic functions* (3.3) and (3.4) by the correction coefficient $\frac{2}{\sqrt{5}}$, hereby turning his *"golden" hyperbolic functions* [28] into *symmetric hyperbolic Fibonacci functions* [75]. Only due to such a transformation, there appear in the Bodnar theory [28] the *Fibonacci numbers*, which are connected with the *symmetric hyperbolic Fibonacci functions* [75] by the formulas (3.5) and (3.6).

3.1.6. A brief history

The discovery of the new class of the hyperbolic functions, based on the *golden ratio* and the *Fibonacci numbers*, without any doubt, is an outstanding achievement in the field of modern mathematics; this mathematical achievement can rightly be called the *"golden" paradigm for the "elementary functions"*.

The discovery of the *"golden hyperbolic functions"* and the *"hyperbolic Fibonacci and Lucas functions"* was made about the same time (late 1980s) independent of each other by the Ukrainian scientists Oleg Bodnar [28], on the one hand, and Alexey Stakhov and Ivan Tkachenko [64], on the other hand.

Further development of this theory (introduction of the *symmetric hyperbolic Fibonacci and Lucas functions*) is given in

Stakhov and Rozin's 2005 article [75]. However, the approaches of these scientists to the scientific results were different. Stakhov, Tkachenko and Rozin used a strictly mathematical approach to the new class of hyperbolic functions [64, 75]: for this purpose, they used Binet's formulas (2.33) and (2.34), which connect the Fibonacci and Lucas numbers with the *golden ratio*. Bodnar arrived at this result due to his brilliant scientific intuition and logic of the scientific study of phyllotaxis which led him to the conclusion that the *"golden" hyperbolic functions* (3.3) and (3.4) should reflect the essence of the phenomenon of phyllotaxis and that they should be the basis for the geometric theory of this unique botanical phenomenon.

In any case, the priority in the discovery of the new class of hyperbolic functions belongs to the representatives of the Ukrainian science (Oleg Bodnar, Alexey Stakhov, Ivan Tkachenko, Boris Rozin).

3.1.7. The main Jewish religious symbol is the Shofar

According to Wikipedia (https://en.wikipedia.org/wiki/Shofar), *"The **Shofar** is an ancient musical horn typically made of a ram's horn, used for Jewish religious purposes... The Shofar has been sounded as a sign of victory and celebration. Jewish elders were photographed blowing multiple Shofars after hearing that the Nazis surrendered on May 8, 1945. The Shofar has played a major role in the pro-Israel movement and often played in the Salute to Israel Parade and other pro-Israel demonstrations."*

The idea of writing the article [77] belongs to Boris Rozin, the talented graduate of the Computer Engineering Department of Vinnitsa Technical University. Doctor of Technical Sciences, Professor Alexey Stakhov, who headed this department during Boris Rozin's student years, supported the idea of Boris Rosin to write the article [77], dedicated to this most famous Jewish religious symbol. The article [77] is a continuation of the most famous article by Stakhov and Rozin [75]. Both articles [75, 77] evoked great interest in various countries (especially in Israel) and are the most cited articles written by Boris Rozin together with Alexey Stakhov (Fig. 3.4).

Fig. 3.4. Shofar, the Jewish religious symbol.

3.1.8. Quasi-sinusoidal Fibonacci and Lucas functions

Let's consider the Binet formulas for the Fibonacci and Lucas numbers presented in the following form:

$$F_n = \frac{\Phi^n - (-1)^n \Phi^{-n}}{\sqrt{5}}, \tag{3.9}$$

$$L_n = \Phi^n + (-1)^n \Phi^{-n}, \tag{3.10}$$

where $\Phi = \frac{1+\sqrt{5}}{2}$ is the golden proportion, $n = 0, \pm 1, \pm 2, \pm 3, \dots$.

By comparing the Binet formulas (3.9) and (3.10) with the symmetric hyperbolic functions (2.35)–(2.38), we can see that the continuous functions Φ^x and Φ^{-x} in the formulas (2.35)–(2.38) correspond to the "discrete" sequences Φ^n and Φ^{-n} in the Binet formulas (3.9) and (3.10).

Let's consider the trigonometric function $\cos(\pi x)$, which turns into the "discrete" function $\cos(\pi n)$, if the "continuous" variable x in the function $\cos(\pi x)$ is replaced by the discrete variable $n = 0, \pm 1, \pm 2, \pm 3, \dots$. It is clear that the "discrete" function $\cos(\pi n)$ takes the alternative values (-1) and $(+1)$ in the "discrete" points $n = 0, \pm 1, \pm 2, \pm 3, \dots$.

These arguments, carried out in the article by Stakhov and Rozin [77], became the basis for the introduction of the new continuous function related to the Fibonacci and Lucas numbers.

Definition 3.1. The following continuous function is called the *quasi-sinusoidal Fibonacci function*:

$$Q_F(x) = \frac{\Phi^x - \cos(\pi x)\Phi^{-x}}{\sqrt{5}}. \tag{3.11}$$

There exists the following relationship between the *Fibonacci numbers* F_n, given by (3.9), and *quasi-sinusoidal Fibonacci function*, given by (3.11):

$$F_n = Q_F(n) = \frac{\Phi^n - \cos(\pi n)\Phi^{-n}}{\sqrt{5}}, \tag{3.12}$$

where $n = 0, \pm 1, \pm 2, \pm 3, \ldots$.

Definition 3.2. The following continuous function is called the *quasi-sinusoidal Lucas function*:

$$Q_L(x) = \Phi^x + \cos(\pi x)\Phi^{-x}. \tag{3.13}$$

There exists the following relationship between the *Lucas numbers* L_n, given by (3.10), and the *quasi-sinusoidal Lucas function*, given by (3.13):

$$L_n = Q_L(n) = \Phi^n + \cos(\pi n)\Phi^{-n}. \tag{3.14}$$

3.1.9. Graphs of the quasi-sinusoidal Fibonacci and Lucas functions

The graph of the *quasi-sinusoidal Fibonacci function* (3.11) is a *quasi-sinusoidal curve*, which passes through all the points, corresponding to the *Fibonacci numbers*, given by (3.9), on the coordinate plane (Fig. 3.5).

As shown by the dash line in Fig. 3.5, the symmetric hyperbolic Fibonacci functions (2.35) and (2.36) *envelop* the quasi-sinusoidal Fibonacci function (3.11).

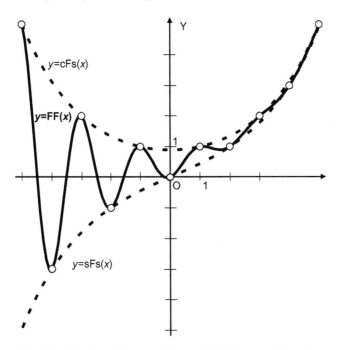

Fig. 3.5. Graph of the quasi-sinusoidal Fibonacci function.

The graph of the *quasi-sinusoidal Lucas function* (3.13) is a *quasi-sinusoidal curve*, which passes through all the points, corresponding to the *Lucas numbers*, given by (3.14), on the coordinate plane (Fig. 3.6). As shown by the dash line in Fig. 3.6, the *symmetric hyperbolic Lucas functions* (2.37) and (2.38) *envelop* the *quasi-sinusoidal Lucas function* (3.14).

The symmetric hyperbolic Lucas functions (2.37) and (2.38) (Fig. 3.6) are the envelopes of the *quasi-sinusoidal Lucas function* (3.13).

3.1.10. Recurrent properties of the quasi-sinusoidal Fibonacci and Lucas functions

Let's now formulate some theorems for the *quasi-sinusoidal Fibonacci and Lucas functions*, proved in Ref. [77].

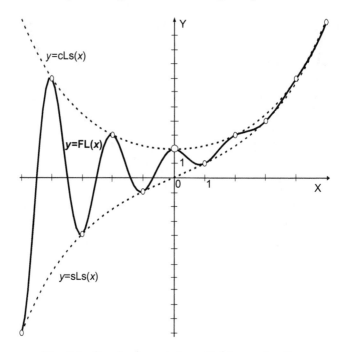

Fig. 3.6. Graph of quasi-sinusoidal Lucas function.

Theorem 3.1. *For the quasi-sinusoidal Fibonacci function, there is a relation similar to the recurrent relation for the Fibonacci numbers* $F_{n+2} = F_{n+1} + F_n$:

$$Q_F(x+2) = Q_F(x+1) + Q_F(x). \qquad (3.15)$$

Theorem 3.2. *For the quasi-sinusoidal Lucas function, there is a relation similar to the recurrent relation for the Lucas numbers* $L_{n+2} = L_{n+1} + L_n$:

$$Q_L(x+2) = Q_L(x+1) + Q_L(x). \qquad (3.16)$$

Theorem 3.3. *For the quasi-sinusoidal Fibonacci function, there is a relation similar to Cassini's formula* $F_n^2 - F_{n+1}F_{n-1} = (-1)^{n+1}$:

$$[Q_F(x)]^2 - Q_F(x+1)Q_F(x-1) = -\cos(\pi x). \qquad (3.17)$$

3.1.11. Fibonacci 3D spiral

It is well known that trigonometric *sine* and *cosine* can be defined as the horizontal projections of the translational motion of a point on the surface of the infinite rotating cylinder with the radius of 1 and an axis of symmetry that coincides with the axis OX. Such a three-dimensional spiral is described by the complex function $f(x) = \cos x + i \sin x$, where $i = \sqrt{-1}$.

By analogy with such a three-dimensional spiral, the *three-dimensional Fibonacci spiral* was introduced in [77]:

$$S_F(x) = \frac{\Phi^x - \cos(\pi x)\Phi^{-x}}{\sqrt{5}} + i\frac{\sin(\pi x)\Phi^{-x}}{\sqrt{5}}. \qquad (3.18)$$

As shown in Fig. 3.7, this function, by its form, looks like the *spiral*, resembling the crater with a curved end.

The next properties of the function (3.18) had been proved in Ref. [77].

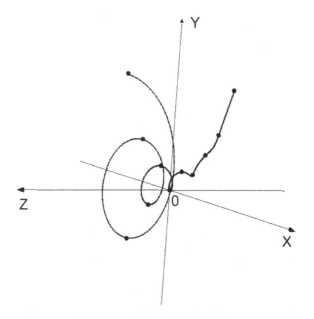

Fig. 3.7. Fibonacci 3D Spiral.

Theorem 3.4. *For the three-dimensional Fibonacci spiral, the following property is true similar to the recurrent relation for the Fibonacci numbers $F_{n+2} = F_{n+1} + F_n$:*

$$S_F(x+2) = S_F(x+1) + S_F(x). \tag{3.19}$$

3.2. The Golden Shofar

The above considerations became the basis for the introduction of the original elementary function called the *Golden Shofar* in Ref. [77].

Denote the real and imaginary parts of the three-dimensional *Fibonacci spiral* (3.18), respectively, by $y(x)$ and $z(x)$, that is,

$$y(x) = \text{Re}[S_F(x)] = \frac{\Phi^x - \cos(\pi x)\Phi^{-x}}{\sqrt{5}} = \frac{\Phi^x}{\sqrt{5}} - \frac{\cos(\pi x)\Phi^{-x}}{\sqrt{5}},$$

$$\tag{3.20}$$

$$z(x) = \text{Im}[S_F(x)] = \frac{\sin(\pi x)\Phi^{-x}}{\sqrt{5}}. \tag{3.21}$$

By using (3.20) and (3.21), we can form the following system of equations:

$$\begin{cases} y(x) - \dfrac{\Phi^x}{\sqrt{5}} = -\dfrac{\cos(\pi x)\Phi^{-x}}{\sqrt{5}}, \\ z(x) = \dfrac{\sin(\pi x)\Phi^{-x}}{\sqrt{5}}. \end{cases} \tag{3.22}$$

Let's square both equations of the system (3.22) and then sum them. Taking y and z as independent variables, we can get some function:

$$\left(y - \frac{\Phi^x}{\sqrt{5}}\right)^2 + z^2 = \left(\frac{\Phi^{-x}}{\sqrt{5}}\right)^2. \tag{3.23}$$

In the three-dimensional space, the function (3.23) corresponds to the curvilinear surface of the second order called the *Golden Shofar* in Ref. [77] (Fig. 3.8).

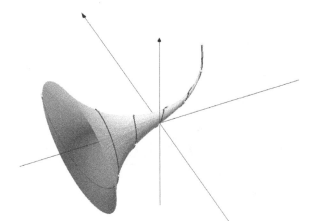

Fig. 3.8. The Golden Shofar.

The resultant three-dimensional surface looks like a horn or crater with a narrow end. Translated from Hebrew, the word "shofar" means a horn, which is a symbol of power and might.

It is easy to show Ref. [77] that the function (3.23) is directly connected with the *symmetric hyperbolic Fibonacci functions* (2.35) and (2.36). By doing some simple transformations [77], we can present the function of the "Golden Shofar" (3.23), as follows:

$$z^2 = [\text{cFs}(x) - y][\text{sFs}(x) + y], \qquad (3.24)$$

where sFs(x) is the *symmetric hyperbolic Fibonacci sine* (2.35) and cFs(x) is the *symmetric hyperbolic Fibonacci cosine* (2.36).

The projection of the "Golden Shofar" onto the XOY is shown in Fig. 3.9. The projection is shown as a dash curve between the plots of the symmetric hyperbolic sine and Fibonacci cosine.

The graph of the function (3.23) "lies" on the *Golden Shofar* and penetrates the plane XOY at the points corresponding to the *Fibonacci numbers* (Fig. 3.9).

The projection of the *Golden Shofar* on the plane XOZ is shown in Fig. 3.10. The projection is shown with a dash curve between two exponential functions, $\frac{-\Phi^{-x}}{\sqrt{5}}$ and $\frac{\Phi^{-x}}{\sqrt{5}}$.

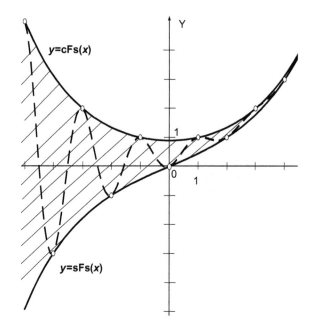

Fig. 3.9. Projection of the Golden Shofar on the plane XOY.

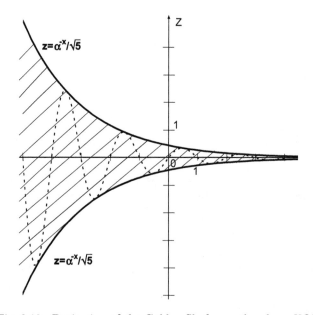

Fig. 3.10. Projection of the Golden Shofar on the plane XOZ.

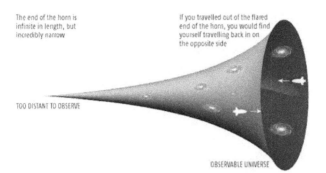

Fig. 3.11. Form of the Universe.

3.3. The Shofar-Like Model of the Universe

3.3.1. Hyperbolic Universes with horned topology

In 2004, a highly authoritative international journal *Classical Quantum Gravity* published a sensational article in the field of cosmology. The article is entitled "Hyperbolic Universes with a Horned Topology and the CMB Anisotropy" (the authors are Ralf Aurich, Sven Lustig, Frank Steiner, Holger Then). The analysis of the latest astrophysical data, especially data on the cosmic microwave background (relic radiation), a kind of "snapshot" of the Universe, which was only 380 000 years old, allowed one to derive equations that determine the curvature and topology of the Universe on a large scale.

These equations led to a sensational conclusion that the Universe in its form is similar to the *horn* (Fig. 3.11).

By comparing Figs. 3.8 and 3.11, it is not difficult to notice the great similarity of the "Golden Shofar" surface, obtained by studying the *Fibonacci hyperbolic functions*, with the shape of the Universe, based on experimental data. Hence, the hypothesis that the Universe is "shofar-like" in its form is quite reasonable. Figure 3.12 shows the "shofar-like topology", which, perhaps, lies at the basis of the model of the hyperbolic Universe.

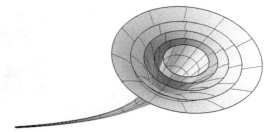

Fig. 3.12. "Shofar-like" topology.

Of course, this is just a hypothesis, but it deserves attention, and its development may lead researchers to new views on the structure of the Universe.

Chapter 4

Theory of Fibonacci and Lucas λ-numbers and its Applications

4.1. Definition of Fibonacci and Lucas λ-numbers

4.1.1. A brief history

The modern *Mathematics of Harmony* [6] is an actively developing area of modern science and mathematics. During the late 20th and early 21st centuries, several researchers from different countries, Argentinean mathematician Vera Spinadel [33], French mathematician of Egyptian origin Midhat Gazale [34], Russian researcher Victor Shenyagin [143], Ukrainian physicist Nikolay Kosinov [144], Ukrainian-Canadian researcher Alexey Stakhov [6, 52, 82], Spanish mathematicians Falcon Sergio, Plaza Angel [145] and others independent of each other began to study a new class of the recurrent numerical sequences that are a generalization of the classical Fibonacci and Lucas numbers.

These numerical sequences, called the *Fibonacci λ-numbers* [52, 82], led to the discovery of the new class of mathematical constants, called by the Argentinean mathematician Vera Spinadel as the *metallic means* or the *metallic proportions* [33]. The number of *metallic proportions* is theoretically infinite; the *classical golden proportion* is their particular case ($\lambda = 1$).

In the book by Midhat Gazale [34], and later in the article by Alexey Stakhov [82], the so-called *Gazale formulas* were introduced, which define the analytical expression for the Fibonacci and Lucas λ-numbers in terms of the *metallic proportions*. By using the *Gazale formulas*, Alexey Stakhov in 2006 introduced the *hyperbolic Fibonacci and Lucas λ-functions* [82]. By using these functions, Alexey Stakhov and Samuil Aranson obtain the original solution of the Hilbert 4th Problem [87–90].

The interest of many researchers from different countries (USA, Argentina, France, Russia, Armenia, Ukraine) in the Fibonacci and Lucas λ-numbers cannot be accidental. This means that the *problem has matured* and scientists from the different countries began to study this problem independent of each other.

This chapter is a popular introduction to the theory of the *Fibonacci and Lucas λ-numbers* and the *hyperbolic Fibonacci and Lucas λ-functions*. This theory leads to the introduction of new mathematical constants, which are a generalization of the classical "golden proportion", and the new "elementary functions", which are a generalization of the above-considered *Fibonacci and Lucas hyperbolic functions*. These mathematical results are of fundamental interest both for mathematics in general (in particular, for the *hyperbolic geometry*) and for all the theoretical natural sciences.

4.1.2. Recurrent relation for the Fibonacci λ-numbers

Let's set a real number $\lambda > 0$ and consider the following recurrent relation:

$$F_\lambda(n+2) = \lambda F_\lambda(n+1) + F_\lambda(n); \quad F_\lambda(0) = F_\lambda(1) = 1, \quad (4.1)$$

where $\lambda > 0$ is arbitrary real number .

The recurrent relation (4.1) "generates" an infinite number of new numerical sequences because each real number $\lambda > 0$ "generates" its own numerical sequence. It is important to emphasize that the particular cases of (4.1) are certain numerical sequences that are widely known in modern science.

In particular, for the case $\lambda = 1$, the recurrent relation (4.1) reduces to the following recurrent relation:

$$F_1(n+2) = F_1(n+1) + F_1(n); \quad F_1(0) = 0, \ F_1(1) = 1, \quad (4.2)$$

which "generates" the classical Fibonacci numbers: $0, 1, 1, 2, 3, 5, 8, 13, 21, 34, \ldots$ Based on this fact, the numerical sequences, generated by the recurrent relation (4.1), were called *Fibonacci λ-numbers* [6, 82].

For the case $\lambda = 2$, the recurrent relation (4.1) reduces to the recurrent relation:

$$F_2(n+2) = 2F_2(n+1) + F_2(n); \quad F_2(0) = 0, F_2(1) = 1, \quad (4.3)$$

which "generates" the so-called *Pell numbers*: $0, 1, 2, 5, 12, 29, 70, \ldots$ (see Wikipedia article "Pell number").

For the case $\lambda = 3, 4$, the recurrent relation (4.1) reduces to the following recurrent relations:

$$F_3(n+2) = 3F_3(n+1) + F_3(n); \quad F_3(0) = 0, \ F_3(1) = 1, \quad (4.4)$$

$$F_4(n+2) = 4F_4(n+1) + F_4(n); \quad F_2(0) = 0, \ F_2(1) = 1. \quad (4.5)$$

Note that the quantity of the recurrent relations, which are "generated" by the general recurrent relation (4.1), is theoretically infinite and equal to the quantity of all real numbers $\lambda > 0$.

4.1.3. Extended Fibonacci λ-numbers

The Fibonacci λ-numbers possess many remarkable properties similar to the properties of these classical Fibonacci numbers. It is proved that these Fibonacci numbers as well as the classical Fibonacci numbers may be "extended" toward the negative values of the discrete variable n.

Table 4.1 shows the four extended Fibonacci λ-numbers corresponding to the values $\lambda = 1, 2, 3, 4$.

Table 4.1. Extended Fibonacci λ-numbers ($\lambda = 1, 2, 3, 4$).

n	0	1	2	3	4	5	6	7	8
$F_1(n)$	0	1	1	2	3	5	8	13	21
$F_1(-n)$	0	1	-1	2	-3	5	-8	13	-21
$F_2(n)$	0	1	2	5	12	29	70	169	408
$F_2(-n)$	0	1	-2	5	-12	29	-70	169	-408
$F_3(n)$	0	1	3	10	33	109	360	1189	3927
$F_3(-n)$	0	1	-3	10	-33	109	-360	1199	-3927
$F_4(n)$	0	1	4	17	72	305	1292	5473	23184
$F_4(-n)$	0	1	-4	17	-72	305	-1292	5473	-23184

Table 4.2. Diagonal sums of Pascal's triangle.

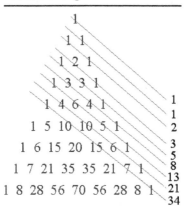

4.2. Representation of the Fibonacci λ-numbers Through Binomial Coefficients

In Vol. I, we studied the so-called *diagonal sums* of Pascal's triangle, which coincide with the classical Fibonacci numbers (Table 4.2)

Let's now represent Pascal's triangle, given by Table 4.2, in the form of the right-angled triangle, as shown in Table 4.3.

The rows (R) of Pascal's triangle will be numbered from top to bottom; the upper row, consisting of the binomial coefficients ($C_0^0 = C_1^0 = C_2^0 = C_3^0 = \cdots = C_n^0 = 1 \cdots$), will be called the *zero-row*. The columns of Pascal's triangle in Table 4.3 will be numbered from left

Table 4.3. Pascal's right-angled triangle.

$\downarrow R/C \rightarrow$	0	1	2	3	$\cdots$	n	$\cdots$
(a) Symbolic form							
0	C_0^0	C_1^0	C_2^0	C_3^0	$\cdots$	C_n^0	$\cdots$
1		C_1^1	C_2^1	C_3^1	$\cdots$	C_3^1	$\cdots$
2			C_2^2	C_3^2	$\cdots$	C_n^2	$\cdots$
3				C_3^3	$\cdots$	C_n^3	$\cdots$
$\vdots$					$\cdots$	$\vdots$	$\cdots$
n					$\cdots$	C_n^n	$\cdots$
2^n	1	2	4	8	$\cdots$	2^n	$\cdots$

$\downarrow R/C \rightarrow$	0	1	2	3	4	5	6
(b) Numerical form							
0	1	1	1	1	1	1	1
1		1	2	3	4	5	6
2			1	3	6	10	15
3				1	4	10	20
4					1	5	15
5						1	6
6							1
Binary numbers	1	2	4	8	16	32	64

to right, and the left column, consisting of only one element ($C_0^0 = 1$), will be called the *zero-column*.

It is known that the sum of the binomial coefficients that form the nth column of Pascal's triangle is equal to the *binary number*:

$$2^n = C_n^0 + C_n^1 + C_n^2 + \cdots + C_n^n. \qquad (4.6)$$

Let's now shift each row of Pascal's triangle in Table 4.3 on the p columns to the right relative to the previous row. As a result, we obtain the table of binomial coefficients, which we call the "*deformed*" Pascal's p-triangle. For the case $p = 1$, the "*deformed*" Pascal's ($p = 1$)-*triangle* is demonstrated by Table 4.4.

If we now sum up the binomial coefficients of the "deformed" Pascal's ($p = 1$)th triangle by one column relative to the previous

Table 4.4. Deformed Pascal's ($p = 1$)-triangle.

↓ R/C →	0	1	2	3	4	5	$\cdots$	2m	2m + 1
(a) Symbolic form									
0	C_0^0	C_1^0	C_2^0	C_3^0	C_4^0	C_5^0	$\cdots$	C_{2m}^0	C_{2m+1}^0
1		C_1^1	C_2^1	C_3^1	C_4^1		$\cdots$	C_{2m-1}^1	C_{2m}^1
2					C_2^2	C_3^2	$\cdots$	C_{2m-2}^2	C_{2m-1}^2
$\vdots$								$\vdots$	$\vdots$
m								C_m^m	C_{m+1}^m

↓ R/C →	0	1	2	3	4	5	6	7	8	9	10
(b) Numerical form											
0	1	1	1	1	1	1	1	1	1	1	1
1			1	2	3	4	5	6	7	8	9
2					1	3	6	10	15	21	28
3							1	4	10	20	35
4									1	5	15
5											1
Fibonacci numbers	1	1	2	3	5	8	13	21	34	55	89

column (see Table 4.4), then we will suddenly arrive at the classical Fibonacci numbers! This unexpected result was first described in the book *Mathematical Discovery* by the American mathematician George Polya [111].

From Table 4.4, it is easy to establish that the sum of binomial coefficients, formed in the nth column of the "deformed" Pascal's triangle, is equal to the Fibonacci number F_{n+1}. Let $n = 2m + r$, where m is a quotient, and r is a reminder of the division n by 2. Then, we have the following representation of the Fibonacci number F_{n+1} as the sum of binomial coefficients:

$$F_{n+1} = C_n^0 + C_{n-1}^1 + C_{n-2}^2 + \cdots + C_{m+r}^m. \qquad (4.7)$$

4.3. Cassini Formula for the Fibonacci λ-numbers

Let's recall that the Cassini formula

$$F_n^2 - F_{n-1}F_{n+1} = (-1)^{n+1}, \qquad (4.8)$$

which links the three neighboring Fibonacci numbers F_{n-1}, F_n, F_{n+1}, is one of the most remarkable identities for the classical Fibonacci numbers.

Alexey Stakhov in his 2006 article [82] and then in his fundamental 2009 book [6] proved that the three neighboring Fibonacci λ-numbers $F_\lambda(n-1), F_\lambda(n), F_\lambda(n+1)$, given by the recurrent relation (4.1), have the following property:

$$F_\lambda^2(n) - F_\lambda(n-1) F_\lambda(n+1) = (-1)^{n+1} \qquad (4.9)$$

for the case $\lambda = 0, \pm 1, \pm 2, \pm 3, \ldots$.

Let's remind that the formula (4.8) was derived in the 17th century by the famous French astronomer Cassini. The formula (4.9) was derived by the author three centuries after the Cassini formula (4.8). The formula (4.9) is the so-called *family attribute*, which unites the *Fibonacci λ-numbers* with the *classical Fibonacci numbers*.

It is clear that the formula (4.9), proved in the 21st century [6.82], is a very broad generalization of Cassini's formula (4.8) proved in the 17th century. In essence, the formula (4.9) is a compact notation of the infinite set of formulas similar to the classical Cassini formula (4.8) for the case $\lambda = 0, \pm 1, \pm 2, \pm 3, \ldots$ and this gives us every right to interpret the formula (4.9) as a *paradigm shift* to the formula Cassini (4.8) for the classical Fibonacci numbers.

4.4. Metallic Proportions by Vera Spinadel

Definition. Let's divide both sides of the recurrent relation (4.1) by $F_\lambda(n+1)$ and represent it in the following form:

$$\frac{F_\lambda(n+2)}{F_\lambda(n+1)} = \lambda + \frac{F_\lambda(n)}{F_\lambda(n+1)} = \lambda + \frac{1}{\frac{F_\lambda(n+1)}{F_\lambda(n)}}. \qquad (4.10)$$

If we denote by x the limit of the relationship $\frac{F_\lambda(n+1)}{F_\lambda(n)}$ for the case $n \to \infty$, then by carrying out the passage to the limit $(n \to \infty)$ in the expression (4.10), we obtain the following quadratic equation:

$$x^2 = \lambda x + 1. \qquad (4.11)$$

We will also use the following traditional form for the quadratic equation (4.11):

$$x^2 - \lambda x - 1 = 0. \tag{4.12}$$

As it is well known, the quadratic equation (4.12) has two roots (the positive x_1 and the negative x_2):

$$x_1 = \frac{\lambda + \sqrt{4 + \lambda^2}}{2}, \tag{4.13}$$

$$x_2 = \frac{\lambda - \sqrt{4 + \lambda^2}}{2}. \tag{4.14}$$

Let's establish some properties of the roots (4.13) and (4.14) by using the so-called *Vieta's formulas* (see Wikipedia article "Vieta's formulas"). According to these formulas, the sum of the roots $x_1 + x_2$ of the quadratic equation (4.12) is equal to the coefficient λ at the variable x:

$$x_1 + x_2 = \lambda, \tag{4.15}$$

and the product of the roots is equal to the free term of equation (4.12):

$$x_1 x_2 = -1. \tag{4.16}$$

If we substitute the roots (4.13) and (4.14) into equation (4.11) instead of x, we get the following two identities for the roots of the equation (4.12):

$$x_1^2 = \lambda x_1 + 1, \tag{4.17}$$

$$x_2^2 = \lambda x_2 + 1. \tag{4.18}$$

If we now multiply or divide repeatedly the right and left parts of the expression (4.17) by x_1 and the expression (4.18) by x_2, we get the following two identities for the powers of the roots x_1 and x_2:

$$x_1^n = \lambda x_1^{n-1} + x_1^{n-2}, \tag{4.19}$$

$$x_2^n = \lambda x_2^{n-1} + x_2^{n-2}, \tag{4.20}$$

where $n = 0, \pm 1, \pm 2, \pm 3, \ldots$.

Denote by Φ_λ the positive root x_1, given by (4.13), and consider a new class of mathematical constants defined by the following expression:

$$\Phi_\lambda = \frac{\lambda + \sqrt{4 + \lambda^2}}{2}. \tag{4.21}$$

Note that the formula (4.21), introduced by Alexey Stakhov in the article [82], plays a "key" role in further reasoning.

First of all, we note that for the case $\lambda = 1$, the formula (4.21) reduces to the formula for the *classical golden proportion*:

$$\Phi = \frac{1 + \sqrt{5}}{2}. \tag{4.22}$$

This fact should attract our special attention to the formula (4.21), which is a generalization of the formula (4.22) for the *classical golden proportion*.

The Argentinean mathematician Vera Spinadel in Ref. [33] called the mathematical constants, defined by the expression (4.21), as the *metallic means* or the *metallic proportions*. If in the expression (4.21), we take $\lambda = 1, 2, 3, 4$, then we get the following mathematical constants, having, according to Spinadel, the following special titles:

$$\Phi_1 = \frac{1 + \sqrt{5}}{2} \quad \text{(the golden proportion, } \lambda = 1\text{)},$$

$$\Phi_2 = 1 + \sqrt{2} \quad \text{(the silver proportion, } \lambda = 2\text{)},$$

$$\Phi_3 = \frac{3 + \sqrt{13}}{2} \quad \text{(the bronze proportion, } \lambda = 3\text{)},$$

$$\Phi_4 = 2 + \sqrt{5} \quad \text{(the copper proportion, } \lambda = 4\text{)}.$$

The remaining metallic proportions ($\lambda \geq 5$) do not have special titles:

$$\Phi_5 = \frac{5 + \sqrt{29}}{2}, \quad \Phi_6 = 3 + 2\sqrt{10}, \quad \Phi_7 = \frac{7 + 2\sqrt{14}}{2}, \quad \Phi_8 = 4 + \sqrt{17}.$$

It is not difficult to prove that the root x_2 can be expressed in terms of the *metallic proportions* (4.21) as follows:

$$x_2 = -\frac{1}{\Phi_\lambda} = \frac{\lambda - \sqrt{4 + \lambda^2}}{2}. \tag{4.23}$$

It is clear that the number of *metallic proportions*, given by (4.21), is theoretically infinite because every real number $\lambda > 0$ "generates" its own *metallic proportion* of the kind (4.21). The most important point is the fact that the *metallic proportions* (4.21) are a generalization of the *golden proportion* (4.22) ($\lambda = 1$), one of the most important mathematical constants of the ancient geometry.

This gives us a right to assume that the "metallic proportions" (4.21) represent themselves as a new class of mathematical constants, which, perhaps, are of the same interest for contemporary mathematics and theoretical natural sciences as the ancient golden ratio.

It is curious to emphasize that the simplest quadratic equation of the kind (4.12) had been known for more than 2000 years (Babylon, ancient Egypt, ancient India, ancient China, and ancient Greece). Nevertheless, it was only at the end of 20th century and the beginning of 21st century that the scientists from different countries (Spinadel, Gazale, Kappraff, Tatarenko, Arakelyan, Shenyagin, Kosinov and others) noticed the uniqueness of the quadratic equation (4.12), the study of which led to the discovery of the new class of mathematical constants, the *metallic proportions* (4.21).

4.5. Representation of the "Metallic Proportions" in Radicals

Taking into consideration that $\Phi_\lambda = x_1$, we can represent the expression (4.17) in the following form:

$$\Phi_\lambda^2 = \lambda\Phi_\lambda + 1. \tag{4.24}$$

From the expression (4.24) there follows directly a number of remarkable properties of the *metallic proportions* (4.21). For example, the expression (4.24) can be represented as follows in the radical form:

$$\Phi_\lambda = \sqrt{1 + \lambda\Phi_\lambda}. \tag{4.25}$$

If now instead of Φ_λ in the *radical expression* (4.25), we substitute the expression for Φ_λ, given by (4.25), then we will get the following representation Φ_λ, which we name the *fractal representation* of the mathematical constant Φ_λ:

$$\Phi_\lambda = \sqrt{1 + \lambda\sqrt{1 + \lambda\Phi_\lambda}}.$$

By continuing the process of such substitution *ad infinitum*, we get the following *fractal representation* of the metallic proportion Φ_λ in radicals valid for the arbitrary $\lambda > 0$:

$$\Phi_\lambda = \sqrt{1 + \lambda\sqrt{1 + \lambda\sqrt{1 + \lambda\sqrt{1 + \cdots}}}}. \qquad (4.26)$$

Note for the case $\lambda = 1$, the expression (4.26) reduces to the following well-known *fractal representation* of the classical golden proportion:

$$\Phi = \sqrt{1 + \sqrt{1 + \sqrt{1 + \sqrt{\cdots}}}}. \qquad (4.27)$$

4.6. Representation of the "Metallic Proportions" in the Form of Chain Fraction

Now, we present the expression (4.24) in the following form:

$$\Phi_\lambda = \lambda + \frac{1}{\Phi_\lambda}. \qquad (4.28)$$

If now instead of Φ_λ in the right part of the expression (4.28), we substitute many times the expression for Φ_λ, given by (4.28), and we direct this process *ad infinitum*, we get another *fractal representation* for Φ_λ in the form of the following chain fraction:

$$\Phi_\lambda = \lambda + \cfrac{1}{\lambda + \cfrac{1}{\lambda + \frac{1}{\lambda + \cdots}}}. \qquad (4.29)$$

Note that for the case $\lambda = 1$, the expression (4.29) reduces to the following well-known representation of the *classical golden ratio* in the chain fractions:

$$\Phi = 1 + \cfrac{1}{1 + \cfrac{1}{1 + \cfrac{1}{1 + \cdots}}}. \tag{4.30}$$

As mentioned in Vol. I, this property highlights the *classical golden ratio* $\Phi = \frac{1+\sqrt{5}}{2}$ among other irrational numbers, that is, the *"golden proportion"* (4.22) is the *"unique"* *mathematical constant*. But the property (4.29) is a generalization of the "unique" property (4.30) and therefore the *metallic proportions*, defined by (4.21), can also be classified as the *"unique"* *mathematical constants*.

Let's now write some additional important identities for the *metallic proportion* Φ_λ as follows:

$$\Phi_\lambda + \frac{1}{\Phi_\lambda} = \sqrt{4 + \lambda^2}, \tag{4.31}$$

$$\Phi_\lambda^n = \lambda \Phi_\lambda^{n-1} + \Phi_\lambda^{n-2}, \tag{4.32}$$

where $n = 0, \pm 1, \pm 2, \pm 3, \ldots$.

Thus, the above reasoning provides the additional confirmation of the fact that the *metallic proportions* (4.21) are really the *new "unique" mathematical constant* similar to the *classical golden ratio*.

It is possible to suggest that the study of the properties of the *metallic proportions* (4.21) and the search for their applications in theoretical natural sciences is one of the most important tasks of modern science.

4.7. Self-similarity Principle and Gazale Formulas

4.7.1. Gazale's formulas for the Fibonacci λ-numbers

The formula (4.1) defines the λ-numbers $F_\lambda(n)$ recursively. However, the λ-numbers $F_\lambda(n)$ can be expressed in analytical form through the *metallic proportions* Φ_λ, just as the *Fibonacci numbers* can be expressed analytically through the classical *golden ratio* Φ by using *Binet's formulas*.

By using the roots x_1 and x_2 of the algebraic equation (4.12), we search the analytical formula for the Fibonacci λ-numbers $F_\lambda(n)$ in the following form:

$$F_\lambda(n) = k_1 x_1^n + k_2 x_2^n, \tag{4.33}$$

where k_1 and k_2 are constant coefficients, which are the solutions of the following system of the algebraic equations:

$$\begin{cases} F_\lambda(0) = k_1 x_1^0 + k_2 x_2^0 = k_1 + k_2, \\ F_\lambda(1) = k_1 x_1^1 + k_2 x_2^1 = k_1 \Phi_\lambda - k_2 \dfrac{1}{\Phi_\lambda}. \end{cases} \tag{4.34}$$

Taking into consideration that $F_\lambda(0) = 0$ and $F_\lambda(1) = 1$, we can rewrite the system (4.34) as follows:

$$\begin{cases} k_1 = -k_2, \\ k_1 \Phi_\lambda + k_1 \dfrac{1}{\Phi_\lambda} = k_1 \left(\Phi_\lambda + \dfrac{1}{\Phi_\lambda} \right) = 1. \end{cases} \tag{4.35}$$

From the system of equations (4.35), it is easy to find the following formulas for the coefficients k_1 and k_2:

$$k_1 = \frac{1}{\sqrt{4 + \lambda^2}}, \quad k_2 = -\frac{1}{\sqrt{4 + \lambda^2}}. \tag{4.36}$$

By using (4.36), we can represent the formula (4.33) as follows:

$$F_\lambda(n) = \frac{1}{\sqrt{4 + \lambda^2}} x_1^n - \frac{1}{\sqrt{4 + \lambda^2}} x_2^n = \frac{1}{\sqrt{4 + \lambda^2}} (x_1^n - x_2^n). \tag{4.37}$$

Taking into consideration that $x_1 = \Phi_\lambda$ and $x_2 = -\frac{1}{\Phi_\lambda}$, we can represent the formula (4.37) as follows:

$$F_\lambda(n) = \frac{\Phi_\lambda^n - (-1/\Phi_\lambda)^n}{\sqrt{4 + \lambda^2}}, \tag{4.38}$$

or

$$F_\lambda(n) = \frac{1}{\sqrt{4 + \lambda^2}} \left[\left(\frac{\lambda + \sqrt{4 + \lambda^2}}{2} \right)^n - \left(\frac{\lambda - \sqrt{4 + \lambda^2}}{2} \right)^n \right]. \tag{4.39}$$

Note that, for the first time, the formula (4.39) was derived by Midhat Gazale in Ref. [34].

Let's consider the special cases of the formula (4.39). For the case $\lambda = 1$, the formula (4.39) reduces to the Binet formula for the classical Fibonacci numbers F_n:

$$F_n = F_1(n) = \frac{1}{\sqrt{5}}\left[\left(\frac{1 + \sqrt{5}}{2}\right)^n - \left(\frac{1 - \sqrt{5}}{2}\right)^n\right]. \qquad (4.40)$$

For the case $\lambda = 2$, the formula (4.39) takes the following form:

$$F_2(n) = \frac{1}{2\sqrt{2}}[(1 + \sqrt{2})^n - (1 - \sqrt{2})^n]. \qquad (4.41)$$

It is believed that the formula (4.41) was first derived by the English mathematician Pell (1610–1685) and this formula "generates" the so-called *Pell numbers* (see Wikipedia article "Pell number").

Taking into consideration the authorship of Midhat Gazale in the proof and the usage of the formula (4.39) [34], Alexey Stakhov in Ref. [82] named the formula (4.39) the *Gazale formula for the Fibonacci λ-numbers*. Note that after simple transformations, the *Gazale formula* (4.39) can be written as follows:

$$F_\lambda(n) = \frac{\Phi_\lambda^n - (-1)^n\Phi_\lambda^{-n}}{\sqrt{4 + \lambda^2}}, \qquad (4.42)$$

where $n = 0, \pm 1, \pm 2, \pm 3, \ldots$.

It is important to emphasize once again that the *Gazale formula* (4.42) for the Fibonacci λ-numbers is a generalization of the well-known Binet formula for the classical Fibonacci numbers. Moreover, the Gazale formula (4.42) "generates" an infinite number of the formulas of the kind (4.42) because each real number $\lambda > 0$ "generates" its own Gazale formula of the kind (4.42).

Let's recall that *self-similarity* is one of the most important concepts of modern mathematics and theoretical natural sciences (see Wikipedia article "Self-similarity"). This notion can be found in mathematics, nature, cybernetics, fine art, and so on. In mathematics, a *self-similar* object is exactly or approximately *similar* to a part of itself (i.e., the whole has the same shape as one or more of the parts). Many objects in the real world, such as coastlines, are statistically *self-similar*: parts of them show the same statistical properties at many scales.

In the Wikipedia article "Patterns in Nature", the following examples of self-similarity in Nature are given:

> "**Patterns in Nature** are visible regularities of form, found in the natural world. These patterns recur in different contexts and can sometimes be modeled mathematically.
>
> Natural patterns include symmetries, trees, spirals, meanders, waves, foams, tessellations, cracks, and stripes.
>
> Early Greek philosophers studied patterns, with Plato, Pythagoras and sat tempting to explain order in nature. The modern understanding of visible patterns developed gradually over time.
>
> In the 19th century, Belgian physicist Joseph Plateau examined soap films, leading him to formulate the concept of a minimal surface. German biologist and artist Ernst Haeckel painted hundreds of marine organisms to emphasize their symmetry. Scottish biologist D'Arcy Thompson pioneered the study of growth patterns in both plants and animals, showing that simple equations could explain spiral growth.
>
> In the 20th century, British mathematician Alan Turing predicted mechanisms of morphogenesis, which give rise to patterns of spots and stripes. Hungarian biologist Aristid Lindenmayer and French American mathematician Benoit Mandelbrot showed how the mathematics of fractals could create plant growth patterns".

In his excellent book [34], Midhat Gazale drew special attention to the connection of the Fibonacci sequences of the order m, given by the recurrent formula $F_{m,n+2} = F_{m,n} + mF_{m,n+1}$ with the concept of *self-similarity*, which, according to Gazale, is the *central concept* of the book [34].

By comparing Gazale's recurrent formula $F_{m,n+2} = F_{m,n} + mF_{m,n+1}$ with the recurrent formula for the Fibonacci λ-numbers (4.1), it is easy to show that for the case $m = \lambda$, these recurrent formulas coincide.

In this regard, we again turn to the remarkable book of Midhat Gazale [34], in which he formulated the following idea:

"A key role in the study of self-similarity is played by numerical sequences, which I call here Fibonaci sequences of the order m, where $F_{m,n+2} = F_{m,n} + mF_{m,n+1}$".

According to Gazale, this recurrent formula expresses the *central concept* of the book [34]: the concept of *self-similarity*. This means that the theory of Fibonacci λ-numbers, outlined in Chapter 4, is nothing more than the mathematical theory of the *self-similarity*, one of the most important concepts of science and living Nature.

4.7.2. Gazale formula for the Lucas λ-numbers

Let's again consider the formula (4.33), which analytically defines the *Fibonacci* λ-numbers through the roots of the algebraic equation (4.12). By analogy with the *Binet formula* for the classical Lucas numbers, we consider the following formula, which, as shown below, defines another interesting numerical sequence [6, 82]:

$$L_\lambda(n) = x_1^n + x_2^n, \tag{4.43}$$

where x_1 and x_2 are the roots of the algebraic equation (4.12).

As stated above, the formula (4.33) in general case ($\lambda > 0$) gives the *Fibonacci* λ-*numbers* (4.1), which for the case $\lambda = 1$ reduces to the classical Fibonacci numbers. By analogy with Lucas numbers, we can assume that, in the general case, the formula (4.43) defines a new class of numerical sequences, which we call the *Lucas* λ-*numbers*.

Let's calculate the initial values for the Lucas λ-*numbers* (4.43), which correspond to the values $n = 0$ and $n = 1$:

$$L_\lambda(0) = x_1^0 + x_2^0 = 1 + 1 = 2, \tag{4.44}$$

$$L_\lambda(1) = x_1^1 + x_2^1 = \lambda. \tag{4.45}$$

By using the identities (4.42) and (4.43), we can represent the formula (4.43) in the following form:

$$\begin{aligned}
L_\lambda(n) = x_1^n + x_2^n &= \lambda x_1^{n-1} + x_2^{n-2} + \lambda x_1^{n-2} + x_2^{n-2} \\
&= \lambda(x_1^{n-1} + x_2^{n-1}) + (x_1^{n-2} + x_2^{n-2}).
\end{aligned} \tag{4.46a}$$

Then, by using (4.43), and also the initial values (4.44) and (4.45), we can represent the formula (4.46a) in the following recurrent form:

$$L_\lambda(n) = \lambda L_\lambda(n-1) + L_\lambda(n-2); \quad L_\lambda(0) = 2, \ L_\lambda(1) = \lambda.$$
(4.46b)

If we substitute in the formula (4.43) x_1 and x_2 with their expressions $x_1 = \Phi_\lambda$ and $x_2 = -\frac{1}{\Phi_\lambda}$, then the expression (4.43) takes the following form:

$$L_\lambda(n) = \left[\Phi_\lambda^n + \left(\frac{-1}{\Phi_\lambda} \right)^n \right].$$
(4.47)

After the simple transformations, the formula (4.47) can be written in the following form:

$$L_\lambda(n) = \Phi_\lambda^n + (-1)^n \Phi_\lambda^{-n}.$$
(4.48)

The formula (4.48) was first derived in Ref. [82] and called the *Gazale formula for Lucas λ-numbers.*

It is important to emphasize that the Gazale formula (4.48) for the Lucas λ-numbers is a generalization of the well-known Binet formula for the classical Lucas numbers. In this case, the Gazale formula (4.48) generates an infinite number of the formulas of the kind (4.48) because each real number $\lambda > 0$ generates its own Gazale formula of the kind (4.48).

4.8. Hyperbolic Fibonacci and Lucas λ-functions

4.8.1. Properties of the extended Fibonacci and Lucas λ-numbers

Let's now investigate the Fibonacci and Lucas λ-numbers extended toward the negative values of the discrete variable n, and study the properties of the "extended" Fibonacci and Lucas λ-numbers. For this purpose, we represent the Gazale formula (4.42) for the Fibonacci λ-numbers for the case of the negative values of n in the following form:

$$F_\lambda(-n) = \frac{\Phi_\lambda^{-n} - (-1)^{-n}\Phi_\lambda^n}{\sqrt{4+\lambda^2}}. \qquad (4.49)$$

By comparing the expressions (4.49) and (4.42), we find that for the *odd* $n = 2k+1$, the Fibonacci λ-numbers $F_\lambda(2k+1)$ and $F_\lambda(-2k-1)$ coincide, but for the *even* $n = 2k$, they are opposite by sign, that is,

$$F_\lambda(2k) = -F_\lambda(-2k) \quad \text{and}$$
$$F_\lambda(2k+1) = F_\lambda(-2k-1). \qquad (4.50)$$

This means that the sequence of the Fibonacci λ-numbers in the range $n = 0, \pm1, \pm2, \pm3, \ldots$ is a symmetric sequence relative to the *Fibonacci λ-number* $F_\lambda(0) = 0$ if we take into consideration the fact that the Fibonacci λ-numbers $F_\lambda(2k)$ and $F_\lambda(-2k)$ are opposite by sign.

Let's carry out similar reasoning for the Lucas λ-numbers. To do this, we represent the formula (4.48) for the negative values of n in the following form:

$$L_\lambda(-n) = \Phi_\lambda^{-n} + (-1)^{-n}\Phi_\lambda^n. \qquad (4.51)$$

By comparing the expressions (4.51) and (4.48) for the *even* $(n = 2k)$ and the *odd* $(n = 2k+1)$ values of n, we can do the following conclusion:

$$L_\lambda(2k) = L_\lambda(-2k) \quad \text{and}$$
$$L_\lambda(2k+1) = -L_\lambda(-2k-1). \qquad (4.52)$$

This means that the Lucas λ-numbers in the range $n = 0, \pm1, \pm2, \pm3, \ldots$ are symmetric sequence relative to the Lucas λ-number $L_\lambda(0) = 2$ if we take into consideration the fact that the Lucas λ-numbers $L_\lambda(2k+1)$ and $L_\lambda(-2k-1)$ are opposite by sign.

The properties (4.50) and (4.52) can be directly observed in the examples given in Table 4.1.

Thus, as a result of the above simple mathematical reasoning, we came to the discovery of the two remarkable mathematical formulas

(4.42) and (4.48), which express in analytical form the two new classes of recurrent numerical sequences, *the Fibonacci and Lucas λ-numbers.*

4.8.2. Definition of the hyperbolic Fibonacci and Lucas λ-functions

The Gazale formulas (4.42) and (4.48) are the initial formulas for the definition of the new class of hyperbolic functions, the *hyperbolic Fibonacci and Lucas λ-functions*, introduced in [82]. Let's consider these functions:

Hyperbolic Fibonacci λ-sine and λ-cosine:

$$
\begin{aligned}
\mathrm{sF}_\lambda(x) &= \frac{\Phi_\lambda^x - \Phi_\lambda^{-x}}{\sqrt{4+\lambda^2}} \\
&= \frac{1}{\sqrt{4+\lambda^2}}\left[\left(\frac{\lambda+\sqrt{4+\lambda^2}}{2}\right)^x - \left(\frac{\lambda+\sqrt{4+\lambda^2}}{2}\right)^{-x}\right],
\end{aligned}
$$
$$(4.53)$$

$$
\begin{aligned}
\mathrm{cF}_\lambda(x) &= \frac{\Phi_\lambda^x + \Phi_\lambda^{-x}}{\sqrt{4+\lambda^2}} \\
&= \frac{1}{\sqrt{4+\lambda^2}}\left[\left(\frac{\lambda+\sqrt{4+\lambda^2}}{2}\right)^x + \left(\frac{\lambda+\sqrt{4+\lambda^2}}{2}\right)^{-x}\right].
\end{aligned}
$$
$$(4.54)$$

Hyperbolic Lucas λ-sine and λ-cosine:

$$
\mathrm{sL}_\lambda(x) = \Phi_\lambda^x - \Phi_\lambda^{-x} = \left(\frac{\lambda+\sqrt{4+\lambda^2}}{2}\right)^x - \left(\frac{\lambda+\sqrt{4+\lambda^2}}{2}\right)^{-x},
$$
$$(4.55)$$

$$
\mathrm{cL}_\lambda(x) = \Phi_\lambda^x + \Phi_\lambda^{-x} = \left(\frac{\lambda+\sqrt{4+\lambda^2}}{2}\right)^x + \left(\frac{\lambda+\sqrt{4+\lambda^2}}{2}\right)^{-x},
$$
$$(4.56)$$

where x is the continuous variable and $\lambda > 0$ is the given positive real number.

The Fibonacci and Lucas λ-numbers are defined through the hyperbolic Fibonacci and Lucas λ-functions as follows:

$$F_\lambda(n) = \begin{cases} \text{sF}_\lambda(n), & n = 2k, \\ \text{cF}_\lambda(n), & n = 2k+1, \end{cases} \tag{4.57}$$

$$L_\lambda(n) = \begin{cases} \text{cL}_\lambda(n), & n = 2k, \\ \text{sL}_\lambda(n), & n = 2k+1. \end{cases} \tag{4.58}$$

It follows from the expression (4.57) that for the *even* values of $n = 2k$, the hyperbolic Fibonacci λ-sine $\text{sF}_\lambda(2k)$ coincides with the Fibonacci λ-number $F_\lambda(2k)$, that is, $\text{sF}_\lambda(2k) = F_\lambda(2k)$, but for the *odd* values of $n = 2k+1$, the hyperbolic Fibonacci λ-cosine $\text{cF}_\lambda(2k+1)$ coincides with the Fibonacci λ-number $F_\lambda(2k+1)$.

In the same time, it follows from the expression (4.58) that for the *odd* values of $n = 2k+1$, the hyperbolic Lucas λ-sine $\text{sL}_\lambda(2k+1)$ coincides with the Lucas λ-number $L_\lambda(2k+1)$, that is, $\text{sL}_\lambda(2k+1) = L_\lambda(2k+1)$, but for the *even* values $n = 2k$, the hyperbolic Lucas λ-cosine $\text{cL}_\lambda(2k)$ coincides with the Lucas λ-number $L_\lambda(2k)$.

It is easy to see that the functions (4.53)–(4.56) relate to each other by the following simple relations:

$$\text{sF}_\lambda(x) = \frac{\text{sL}_\lambda(x)}{\sqrt{4+\lambda^2}}, \quad \text{cF}_\lambda(x) = \frac{\text{cL}_\lambda(x)}{\sqrt{4+\lambda^2}}. \tag{4.59}$$

This means that the hyperbolic Fibonacci λ-functions (4.53) and (4.54) differ from the hyperbolic Lucas λ-functions (4.55) and (4.56) only by the constant coefficient $\frac{1}{\sqrt{4+\lambda^2}}$.

Note that for the case $\lambda = 1$, the hyperbolic Fibonacci and Lucas λ-functions (4.53)–(4.56) reduce to the symmetrical hyperbolic Fibonacci and Lucas functions (3.35)–(3.38), introduced in Ref. [75].

Oleg Bodnar showed [42] that the hyperbolic Fibonacci and Lucas functions, based on the classical golden proportion, are the basis of the botanical phenomenon of phyllotaxis. But then we can assume that the hyperbolic Fibonacci and Lucas λ-functions (4.53)–(4.56) may underlie other physical or biological phenomena.

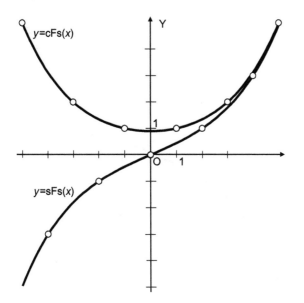

Fig. 4.1. A graph of the symmetric hyperbolic Fibonacci sine and cosine.

4.8.3. Graphs of the hyperbolic Fibonacci and Lucas λ-functions

Graphs of the hyperbolic Fibonacci and Lucas λ-functions have a form similar to the graphs of the symmetrical hyperbolic Fibonacci and Lucas functions.

The difference is in the fact that at the point $x = 0$, the hyperbolic Fibonacci λ-cosine (4.54) takes the value $cF_\lambda(0) = \frac{2}{\sqrt{4+\lambda^2}}$ (see Fig. 4.1), and the hyperbolic Lucas λ-cosine (4.56) takes the value $cL_\lambda(0) = 2$ (see Fig. 4.2).

It is also important to emphasize that the Fibonacci λ-numbers $F_\lambda(n)$ with the *even* values of $n = 0, \pm2, \pm4, \pm6, \ldots$ are "inscribed" into the graph of the hyperbolic Fibonacci λ-sine $sF_\lambda(x)$ in the "discrete" points of the continuous variable $x = 0, \pm2, \pm4, \pm6, \ldots$, while the Fibonacci λ-numbers $F_\lambda(n)$ with the *odd* values $n = \pm1, \pm3, \pm5, \ldots$ are "inscribed" into the graph of the hyperbolic Fibonacci λ-cosine $sF_\lambda(x)$ in the "discrete" points of the continuous variable $x = \pm1, \pm3, \pm5, \ldots$ (see Fig. 4.1).

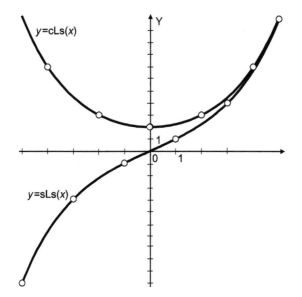

Fig. 4.2. A graph of the symmetric hyperbolic Lucas sine and cosine.

On the other hand, the Lucas λ-numbers $L_\lambda(n)$ with the *even* values $n = 2k$ are "inscribed" into the graph of the hyperbolic Lucas λ-cosine $cL_\lambda(x)$ in the points $x = 0, \pm 2, \pm 4, \pm 6, \ldots$, and the Lucas λ-numbers $L_\lambda(n)$ with the *odd* values $n = 2k + 1$ are "inscribed" into the graph of the hyperbolic Lucas λ-sine $sL_\lambda(x)$ in the "discrete" points of the continuous variable $x = \pm 1, \pm 3, \pm 5, \ldots$. (see Fig. 4.2).

By analogy with the symmetrical hyperbolic Fibonacci and Lucas functions [72], we can introduce other types of the hyperbolic Fibonacci and Lucas λ-functions, in particular, hyperbolic λ-tangents, λ-cotangents, λ-secants and λ-cosecants.

4.9. Special Cases of Hyperbolic Fibonacci and Lucas λ-functions

4.9.1. The "golden", "silver", "bronze" and "copper" hyperbolic Fibonacci and Lucas λ-functions

The formulas (4.53)–(4.56) give infinite number of different hyperbolic Fibonacci and Lucas λ-functions because every real number $\lambda > 0$ "generates" infinite number of different hyperbolic Fibonacci and Lucas λ-functions of the kind (4.53)–(4.56).

Let's consider the characteristic cases of hyperbolic Fibonacci and Lucas λ-functions (4.53)–(4.56) corresponding to the different values of λ.

For the case $\lambda = 1$, the *"golden" proportion* (4.22) is the base of the hyperbolic Fibonacci and Lucas $(\lambda = 1)$-functions, which for the case $\lambda = 1$ coincide with the symmetrical hyperbolic Fibonacci and Lucas functions (2.35)–(2.38). Later on, we will name the functions (2.35)–(2.38) as the *"golden" hyperbolic Fibonacci and Lucas functions*.

For the case $\lambda = 2$, the *"silver"* proportion $\Phi_2 = 1 + \sqrt{2}$ is the base of the hyperbolic Fibonacci and Lucas $(\lambda = 2)$-functions; we will name these functions as the *"silver" hyperbolic Fibonacci and Lucas functions*:

$$\mathrm{sF}_2(x) = \frac{\Phi_2^x - \Phi_2^{-x}}{\sqrt{8}} = \frac{1}{2\sqrt{2}}[(1+\sqrt{2})^x - (1+\sqrt{2})^{-x}], \quad (4.60)$$

$$\mathrm{cF}_2(x) = \frac{\Phi_2^x + \Phi_2^{-x}}{\sqrt{8}} = \frac{1}{2\sqrt{2}}[(1+\sqrt{2})^x + (1+\sqrt{2})^{-x}], \quad (4.61)$$

$$\mathrm{sL}_2(x) = \Phi_2^x - \Phi_2^{-x} = (1+\sqrt{2})^x - (1+\sqrt{2})^{-x}, \quad (4.62)$$

$$\mathrm{cL}_2(x) = \Phi_2^x + \Phi_2^{-x} = (1+\sqrt{2})^x + (1+\sqrt{2})^{-x}. \quad (4.63)$$

For the case $\lambda = 3$, the *"bronze"* proportion $\Phi_3 = \frac{3+\sqrt{13}}{2}$ is the base of a new class of hyperbolic functions; we will name these functions as the *"bronze" hyperbolic Fibonacci and Lucas functions*:

$$\mathrm{sF}_3(x) = \frac{\Phi_3^x - \Phi_3^{-x}}{\sqrt{13}} = \frac{1}{\sqrt{13}}\left[\left(\frac{3+\sqrt{13}}{2}\right)^x - \left(\frac{3+\sqrt{13}}{2}\right)^{-x}\right],$$
$$(4.64)$$

$$\mathrm{cF}_3(x) = \frac{\Phi_3^x + \Phi_3^{-x}}{\sqrt{13}} = \frac{1}{\sqrt{13}}\left[\left(\frac{3+\sqrt{13}}{2}\right)^x + \left(\frac{3+\sqrt{13}}{2}\right)^{-x}\right],$$
$$(4.65)$$

$$\mathrm{sL}_3(x) = \Phi_3^x - \Phi_3^{-x} = \left(\frac{3+\sqrt{13}}{2}\right)^x - \left(\frac{3+\sqrt{13}}{2}\right)^{-x}, \qquad (4.66)$$

$$\mathrm{cL}_3(x) = \Phi_3^x + \Phi_3^{-x} = \left(\frac{3+\sqrt{13}}{2}\right)^x + \left(\frac{3+\sqrt{13}}{2}\right)^{-x}. \qquad (4.67)$$

For the case $\lambda = 4$, the *"copper" proportion* $\Phi_4 = 2+\sqrt{5}$ is the base of a new class of hyperbolic functions; we will name these functions as the *"copper" hyperbolic Fibonacci and Lucas functions*:

$$\mathrm{sF}_4(x) = \frac{\Phi_4^x - \Phi_4^{-x}}{2\sqrt{5}} = \frac{1}{2\sqrt{5}}[(2+\sqrt{5})^x - (2+\sqrt{5})^{-x}], \qquad (4.68)$$

$$\mathrm{cF}_4(x) = \frac{\Phi_4^x + \Phi_4^{-x}}{2\sqrt{5}} = \frac{1}{2\sqrt{5}}[(2+\sqrt{5})^x + (2+\sqrt{5})^{-x}], \qquad (4.69)$$

$$\mathrm{sL}_4(x) = \Phi_4^x - \Phi_4^{-x} = (2+\sqrt{5})^x - (2+\sqrt{5})^{-x}, \qquad (4.70)$$

$$\mathrm{cL}_4(x) = \Phi_4^x + \Phi_4^{-x} = (2+\sqrt{5})^x + (2+\sqrt{5})^{-x}. \qquad (4.71)$$

4.9.2. Connection with the classical hyperbolic functions

Let's now compare the *hyperbolic Lucas λ-functions* (4.55) $\mathrm{sL}_\lambda(x) = \Phi_\lambda^x - \Phi_\lambda^{-x}$ and (4.56) $\mathrm{cL}_\lambda(x) = \Phi_\lambda^x + \Phi_\lambda^{-x}$ to the *classical hyperbolic functions* $\mathrm{sh}(x) = \frac{e^x - e^{-x}}{2}$, $\mathrm{ch}(x) = \frac{e^x + e^{-x}}{2}$.

It is easy to prove that the *hyperbolic Lucas λ-functions* (*sine and cosine*) coincide with the *classical hyperbolic functions* (*sine and cosine*) up to the constant coefficient of $1/2$, that is,

$$\mathrm{sh}(x) = \frac{\mathrm{sL}_\lambda(x)}{2} \quad \text{and} \quad \mathrm{ch}(x) = \frac{\mathrm{cL}_\lambda(x)}{2}. \qquad (4.72)$$

4.9.3. Connection to Pell's numbers

It is well known (see Wikipedia article "Pell number") that the *Pell's number* P_n is defined by the following recurrent relation:

$$P_n = 2P_{n-1} + P_{n-2}, \quad P_0 = 0, P_1 = 1. \qquad (4.73)$$

Table 4.5. "Extended" Pell numbers.

n	0	1	2	3	4	5	6	7	8	9	10
P_n	0	1	2	5	12	29	70	169	408	985	2738
P_{-n}	0	1	-2	5	-12	29	-70	169	-408	985	-2738

The first few *Pell's numbers* look like this: $0, 1, 2, 5, 12, 29,$ $70, 169, 408, 985, 2378, \ldots$ In other words, the sequence of *Pell's numbers* starts with 0 and 1, and each subsequent *Pell's number* P_n is equal to the sum of the doubled previous $2P_{n-1}$ and the Pell's number P_{n-2} standing before it.

It is clear that if the Pell numbers will be "extended" toward the negative values of the index n, then we get a sequence of numbers that coincide with the *Fibonacci λ-numbers*, corresponding to the case $\lambda = 2$, that is, $P_n = F_2(n)$. Table 4.5 provides the "extended" Pell numbers.

By using formula (4.41) $P_n = F_2(n) = \frac{1}{2\sqrt{2}}[(1+\sqrt{2})^n - (1-\sqrt{2})^n]$, the Pell numbers can be represented in the following analytical form:

$$P_n = \frac{1}{2\sqrt{2}}[(1+\sqrt{2})^n - (1-\sqrt{2})^n], \qquad (4.74)$$

where $\Phi_2 = 1 + \sqrt{2}$ is the *"silver" proportion*.

4.10. The Most Important Formulas and Identities for the Hyperbolic Fibonacci and Lucas λ-functions

4.10.1. The relations connecting the "metallic proportions" with the "golden proportion"

We emphasize once again that the number of hyperbolic Fibonacci and Lucas functions, given by (4.53)–(4.56), can be continued *ad infinitum*. It is important to emphasize that they are directly related to some well-known numerical sequences: the *Fibonacci numbers*, the *Lucas numbers*, and the *Pell numbers*. These functions retain all the most important properties of the classical hyperbolic functions [140] and the symmetric hyperbolic Fibonacci and Lucas functions [75] considered above, while they possess, on the one hand, the same *recurrent properties*, similar to the classical Fibonacci and Lucas numbers, on the other hand, they possess the *hyperbolic properties*, similar to the properties of the classical hyperbolic functions [140], which, as shown above, are a special case of Lucas hyperbolic λ-functions.

The bases of these functions are *metallic proportions*, which are a generalization of the ancient *golden proportion*. Let's start with the relations, which connect the ancient *golden proportion* with *metallic proportions*. These relations are given in Table 4.6.

4.10.2. Recurrent properties

Let's consider some of the hyperbolic Fibonacci and Luas functions; they had been proved and described in the papers [59, 74, 75, 77, 79, 80, 82, 84, 92, 94, 97] and in the books [6, 16, 17, 19, 24, 28, 29, 33, 34, 37, 38, 40, 42, 43, 45–47, 49, 50].

Table 4.6. The connection of the classical "golden proportion" with the "metallic proportions".

Golden proportion ($\lambda = 1$)	Metallic proportions ($\lambda > 0$)
$\Phi = \frac{1+\sqrt{5}}{2}$	$\Phi_\lambda = \frac{\lambda+\sqrt{4+\lambda^2}}{2}$
$\Phi = \sqrt{1+\sqrt{1+\sqrt{1+\sqrt{\cdots}}}}$	$\Phi_\lambda = \sqrt{1+\lambda\sqrt{1+\lambda\sqrt{1+\lambda\sqrt{\cdots}}}}$
$\Phi = 1 + \cfrac{1}{1+\cfrac{1}{1+\cfrac{1}{1+\cdots}}}$	$\Phi_\lambda = \lambda + \cfrac{1}{\lambda+\cfrac{1}{\lambda+\cfrac{1}{\lambda+\cdots}}}$
$\Phi^n = \Phi^{n-1} + \Phi^{n-2} = \Phi \times \Phi^{n-1}$	$\Phi_\lambda^n = \lambda\Phi_\lambda^{n-1} + \Phi_\lambda^{n-2} = \Phi_\lambda \times \Phi_\lambda^{n-1}$
$F(n) = \frac{\Phi^n - (-1)^n \Phi^{-n}}{\sqrt{5}}$	$F_\lambda(n) = \frac{\Phi_\lambda^n - (-1)^n \Phi_\lambda^{-n}}{\sqrt{4+\lambda^2}}$
$L(n) = \Phi^n + (-1)^n \Phi^{-n}$	$L_\lambda(n) = \Phi_\lambda^n + (-1)^n \Phi_\lambda^{-n}$
$\mathrm{sFs}(x) = \frac{\Phi^x - \Phi^{-x}}{\sqrt{5}}$	$\mathrm{sF}_\lambda(x) = \frac{\Phi_\lambda^x - \Phi_\lambda^{-x}}{\sqrt{4+\lambda^2}}$
$\mathrm{cFs}(x) = \frac{\Phi^x + \Phi^{-x}}{\sqrt{5}}$	$\mathrm{cF}_\lambda(x) = \frac{\Phi_\lambda^x + \Phi_\lambda^{-x}}{\sqrt{4+\lambda^2}}$
$\mathrm{sLs}(x) = \Phi^x - \Phi^{-x}$	$\mathrm{sL}_\lambda(x) = \Phi_\lambda^x - \Phi_\lambda^{-x}$
$\mathrm{cLs}(x) = \Phi^x + \Phi^{-x}$	$\mathrm{cL}_\lambda(x) = \Phi_\lambda^x + \Phi_\lambda^{-x}$

Let's consider without proof some theorems concerning the recurrent properties of the hyperbolic Fibonacci and Lucas λ-functions.

Theorem 4.1. *The following relations, which are similar to the recurrent relation for the Fibonacci λ-numbers $F_\lambda(n+2) = \lambda F_\lambda(n+1) + F_\lambda(n)$, are true for the hyperbolic Fibonacci λ-functions:*

$$sF_\lambda(x+2) = \lambda cF_\lambda(x+1) + sF_\lambda(x), \tag{4.75}$$

$$cF_\lambda(x+2) = \lambda sF_\lambda(x+1) + cF_\lambda(x). \tag{4.76}$$

Theorem 4.2 (A generalization of Cassini's formula). *The following relations, which are similar to the generalized Cassini's formula for the Fibonacci λ-numbers $F_\lambda^2(n) - F_\lambda(n-1)F_\lambda(n+1) = (-1)^{n+1}$, are true for the hyperbolic Fibonacci λ-sine and λ-cosine:*

$$[sF_\lambda(x)]^2 - cF_\lambda(x+1)\,cF_\lambda(x-1) = -1, \tag{4.77}$$

$$[cF_\lambda(x)]^2 - sF_\lambda(x+1)\,sF_\lambda(x-1) = 1. \tag{4.78}$$

Note that Theorems 4.1 and 4.2 are the examples of the so-called *recurrent properties* of the hyperbolic Fibonacci and Lucas λ-functions.

4.10.3. Hyperbolic properties

Let's now consider some hyperbolic properties of the Fibonacci and Lucas λ-functions (4.53)–(4.56).

Theorem 4.3 (Parity property). *The hyperbolic Fibonacci and Lucas λ-sines are the odd functions, but the hyperbolic Fibonacci and Lucas λ-cosines are the even functions:*

$$sF_\lambda(-x) = -sF_\lambda(x), \quad cF_\lambda(-x) = cF_\lambda(x), \tag{4.79}$$

$$sL_\lambda(-x) = -sL_\lambda(x), \quad cL_\lambda(-x) = cL_\lambda(x). \tag{4.80}$$

Theorem 4.4. *The following identities, similar to the identity $[ch(x)]^2 - [sh(x)]^2 = 1$ for the classical hyperbolic functions, are true* *for the hyperbolic Fibonacci and Lucas λ-functions*:

$$[cF_\lambda(x)]^2 - [sF_\lambda(x)]^2 = \frac{4}{4 + \lambda^2}, \tag{4.81}$$

$$[cL_\lambda(x)]^2 - [sL_\lambda(x)]^2 = 4. \tag{4.82}$$

Theorem 4.5. *The following identity, similar to the identity* $ch(x + y) = ch(x)ch(y) + sh(x)sh(y)$ *for the classical hyperbolic functions, is true for the hyperbolic Fibonacci λ-functions*:

$$\frac{2}{\sqrt{4 + \lambda^2}}cF_\lambda(x + y) = cF_\lambda(x)\,cF_\lambda(y) + sF_\lambda(x)\,sF_\lambda(y). \tag{4.83}$$

Theorem 4.6. *The following identity, similar to the identity* $ch(x - y) = ch(x)ch(y) - sh(x)sh(y)$ *for the classical hyperbolic functions, is true for the hyperbolic Fibonacci λ-functions*:

$$\frac{2}{\sqrt{4 + \lambda^2}}cF_\lambda(x - y) = cF_\lambda(x)\,cF_\lambda(y) - sF_\lambda(x)\,sF_\lambda(y). \tag{4.84}$$

Theorem 4.7. *The following identity, similar to the identity* $ch(2x) = [ch(x)]^2 + [sh(x)]^2$ *for the classical hyperbolic functions, is true for the hyperbolic Fibonacci and Lucas λ-functions*:

$$\frac{2}{\sqrt{4 + \lambda^2}}cF_\lambda(2x) = [cF_\lambda(x)]^2 + [sF_\lambda(x)]^2, \tag{4.85}$$

$$2cL_\lambda(2x) = [cL_\lambda(x)]^2 + [sL_\lambda(x)]^2. \tag{4.86}$$

Theorem 4.8. *The following identity, similar to the identity* $sh(2x) = 2\,sh(x)ch(x)$ *for the classical hyperbolic functions, is true for the hyperbolic Fibonacci and Lucas λ-functions*:

$$\frac{1}{\sqrt{4 + m^2}}sF_\lambda(2x) = sF_\lambda(x)cF_\lambda(x), \tag{4.87}$$

$$sL_\lambda(2x) = sL_\lambda(x)cL_\lambda(x). \tag{4.88}$$

Theorem 4.9. *The following formulas, similar to the Moivre formulas for classical hyperbolic functions* $[ch(x) \pm sh(x)]^n = ch(nx) \pm sh(nx)$, *are true for the hyperbolic Fibonacci and Lucas*

Table 4.7. Hyperbolic properties for the Fibonacci λ-functions.

Formulas for the classical hyperbolic functions	Formulas for hyperbolic Fibonacci λ-functions
$\text{sh}(x) = \frac{e^x - e^{-x}}{2}; \text{ch}(x) = \frac{e^x + e^{-x}}{2}$	$\text{sF}_\lambda(x) = \frac{\Phi_\lambda^x - \Phi_\lambda^{-x}}{\sqrt{4+\lambda^2}}; \text{cF}_\lambda(x) = \frac{\Phi_\lambda^x + \Phi_\lambda^{-x}}{\sqrt{4+\lambda^2}}$
$\text{sh}(x+2) = 2\,\text{sh}(1)\,\text{ch}(x+1) + \text{sh}(x)$	$\text{sF}_\lambda(x+2) = \lambda\,\text{cF}_\lambda(x+1) + \text{sF}_\lambda(x)$
$\text{ch}(x+2) = 2\,\text{sh}(1)\,\text{sh}(x+1) + \text{ch}(x)$	$\text{cF}_\lambda(x+2) = \lambda\,\text{sF}_\lambda(x+1) + \text{cF}_\lambda(x)$
$\text{sh}^2(x) - \text{ch}(x+1)\,\text{ch}(x-1) = -\text{ch}^2(1)$	$[\text{sF}_\lambda(x)]^2 - \text{cF}_\lambda(x+1)\,\text{cF}_\lambda(x-1) = -1$
$\text{ch}^2(x) - \text{sh}(x+1)\,\text{sh}(x-1) = \text{ch}^2(1)$	$[\text{cF}_\lambda(x)]^2 - \text{sF}_\lambda(x+1)\,\text{sF}_\lambda(x-1) = 1$
$\text{ch}^2(x) - \text{sh}^2(x) = 1$	$[\text{cF}_\lambda(x)]^2 - [\text{sF}_\lambda(x)]^2 = \frac{4}{4+\lambda^2}$
$\text{sh}(x+y) = \text{sh}(x)\,\text{ch}(x) + \text{ch}(x)\,\text{sh}(x)$	$\frac{2}{\sqrt{4+\lambda^2}}\text{sF}_\lambda(x+y) = \text{sF}_\lambda(x)\,\text{cF}_\lambda(x) + \text{cF}_\lambda(x)\,\text{sF}_\lambda(x)$
$\text{sh}(x-y) = \text{sh}(x)\,\text{ch}(x) - \text{ch}(x)\,\text{sh}(x)$	$\frac{2}{\sqrt{4+\lambda^2}}\text{sF}_\lambda(x-y) = \text{sF}_\lambda(x)\,\text{cF}_\lambda(x) - \text{cF}_\lambda(x)\,\text{sF}_\lambda(x)$
$\text{ch}(x+y) = \text{ch}(x)\,\text{ch}(x) + \text{sh}(x)\,\text{sh}(x)$	$\frac{2}{\sqrt{4+\lambda^2}}\text{cF}_\lambda(x+y) = \text{cF}_\lambda(x)\,\text{cF}_\lambda(x) + \text{sF}_\lambda(x)\,\text{sF}_\lambda(x)$
$\text{ch}(x-y) = \text{ch}(x)\,\text{ch}(x) - \text{sh}(x)\,\text{sh}(x)$	$\frac{2}{\sqrt{4+\lambda^2}}\text{cF}_\lambda(x-y) = \text{cF}_\lambda(x)\,\text{cF}_\lambda(x) - \text{sF}_\lambda(x)\,\text{sF}_\lambda(x)$
$\text{ch}(2x) = 2\,\text{sh}(x)\,\text{ch}(x)$	$\frac{1}{\sqrt{4+\lambda^2}}\text{cF}_\lambda(2x) = \text{sF}_\lambda(x)\,\text{cF}_\lambda(x)$
$[\text{ch}(x) \pm \text{sh}(x)]^n = \text{ch}(nx) \pm \text{sh}(nx)$	$[\text{cF}_\lambda(x) \pm \text{sF}_\lambda(x)]^n = \left(\frac{2}{\sqrt{4+\lambda^2}}\right)^{n-1}[\text{cF}_\lambda(nx) \pm \text{sF}_\lambda(nx)]$

λ-functions:

$$[\mathrm{cF}_\lambda(x) \pm \mathrm{sF}_\lambda(x)]^n = \left(\frac{2}{\sqrt{4+m^2}}\right)^{n-1} [\mathrm{cF}_\lambda(nx) \pm \mathrm{sF}_\lambda(nx)],$$

$$(4.89)$$

$$[\mathrm{cL}_\lambda(x) \pm \mathrm{sL}_\lambda(x)]^n = 2^{n-1}[\mathrm{cL}_\lambda(nx) \pm \mathrm{sL}_\lambda(nx)]. \tag{4.90}$$

Table 4.7 shows the basic formulas and identities, connecting the hyperbolic Fibonacci λ-functions with the classical hyperbolic functions.

Note that the list of formulas for the hyperbolic Lucas λ-functions can be easily obtained from Table 4.7 if we use the relations (4.59), connecting the hyperbolic Lucas functions with the hyperbolic Fibonacci functions.

Chapter 5

Hilbert Problems: General Information

5.1. A History of the Hilbert Problems [146–149]

From 6 to 12 August, 1900, the *International Congress of Mathematicians* was held in Paris, where the eminent German mathematician David Hilbert (1862–1943) delivered his famous lecture *Mathematical Problems*. In this lecture, Hilbert proposed 23 problems of mathematics, the study of which, in his opinion, can significantly stimulate further development of mathematics and science in general. The ideas having relation to the content of these problems have not lost their relevance until now.

Currently, 16 of 23 problems are solved. Two problems are recognized as incorrect. From the remaining five problems, two problems are not solved, and three problems had been solved only for some partial cases (for more details, see [146–149]).

Next, we focus our attention on solving the two important mathematical *Hilbert Problems* directly related to the *Mathematics of Harmony*. Those are the *10th Problem*, the insolvability of which was proved in 1970 by Yuri Matiyasevich (born in 1947) [150], and the Fourth Problem, the insolvability of which for hyperbolic geometries was proved in 2019 by Alexey Stakhov and Samuil Aranson [151].

5.2. Original Solution of Hilbert's Fourth Problem Based on the Hyperbolic Fibonacci and Lucas λ-Functions

Naturally, Hilbert could not pass by the unsolved mathematical problems associated with non-Euclidean geometry. Hilbert names *Lobachevsky geometry* (the *hyperbolic geometry*) and *Riemann geometry* (the *elliptic geometry*) as the geometries that are closest to *Euclidean geometry*. Hilbert formulates the *Fourth Problem* as follows:

> *"A more general question that arises in this case is the following: is it possible from other fruitful points of view to construct geometries, which could be considered as the closest to ordinary Euclidean geometry?"*

A detailed analysis of all attempts to solve *Hilbert's Fourth Problem* is given in the article "Once again on the Hilbert 4th Problem" [157], the author of which is the well-known Russian mathematician Samuil Aranson, doctor of Physical–Mathematical sciences. Aranson stresses that many mathematicians were involved in solving *Hilbert's Fourth Problem*.

The history of the issue about scientific results relating to the *Hilbert Fourth Problem* is set forth in detail both in the 1969 book *Hilbert Problems* (general editor, academician Aleksandrov) [146]. In this book, we can read the comments to *Hilbert's Fourth Problem* written by the prominent Russian mathematician Yaglom. Besides, we can read similar comments to *Hilbert's Fourth Problem* written by the famous American geometer Busemann and published in Ref. [158].

The first contribution to the solution of this problem is the PhD thesis of the German mathematician Hammel, defended in 1901 under Hilbert's guidance. Hammel's results and comments to them can be found in the Busemann article [158]. As emphasized in Busemann's article, *"Hammel's work, of course, douse not exhaust everything that can be said about Hilbert's Fourth problem; other approaches to which were repeatedly proposed later."*

Apparently, the greatest contribution to the solution of *Hilbert's Fourth Problem* was made by the outstanding Soviet mathematician A.V. Pogorelov, the author of the book, *The Hilbert Fourth Problem* [160].

Unfortunately, the world mathematical community did not perceive Pogorelov's solution, despite the fact that the book was translated into English.

Despite the critical attitude of mathematicians to Hilbert's Fourth Problem, it is necessary to emphasize its extreme importance for the development of mathematics, in particular, geometry. Therefore, the study of this problem continues in modern science. This is why the solution of Hilbert's Fourth problem, described by Stakhov and Aranson in Refs. [68–70], can be considered as a new contribution to the solution of this complex mathematical problem.

5.3. The "Golden" Non-Euclidean Geometry

In 2016, World Scientific published Stakhov and Aranson's book The "Golden" Non-Euclidean Geometry. Hilbert's Fourth Problem, "Golden" Dynamical Systems, and Fine-Structure Constant (see Fig. 5.1).

This unique book overturns our ideas about non-Euclidean geometry and the fine-structure constant, and attempts to solve long-standing mathematical problems. It describes a general theory of "recursive" hyperbolic functions based on the "Mathematics of Harmony," and the "golden," "silver," and other "metallic" proportions. Then, these theories are used to derive an original solution to Hilbert's Fourth Problem for hyperbolic and spherical geometries. On this journey, the book describes the "golden" qualitative theory of dynamical systems based on "metallic" proportions. Finally, it presents a solution to a Millennium Problem by developing the Fibonacci special theory of relativity as an original physical–mathematical solution for the fine-structure constant. It is intended for a wide audience of readers, who are interested in the history

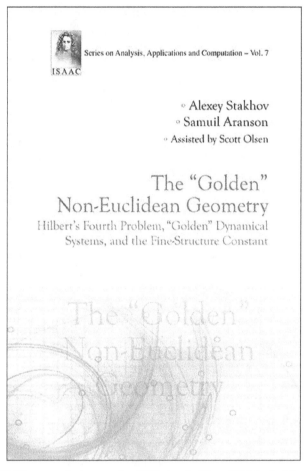

Series on Analysis, Applications and Computation – Vol. 7

ISAAC

° Alexey Stakhov
° Samuil Aranson
° Assisted by Scott Olsen

The "Golden"
Non-Euclidean Geometry
Hilbert's Fourth Problem, "Golden" Dynamical
Systems, and the Fine-Structure Constant

Fig. 5.1. Stakhov and Aranson's 2017 book The "Golden" Non-Euclidean Geometry.

of mathematics, non-Euclidean geometry, Hilbert's mathematical problems, dynamical systems, and Millennium Problems.

5.3.1. Slavic "Golden" Group, International Club of the Golden Section, and Institute of the Golden Section

The first decade of the 21st century was marked by the unprecedented increase in interest toward the ancient problems of *Harmony*,

Platonic solids, and the "*golden section*". In the post-Soviet times, this scientific direction began developing intensively thanks to the activities of the so-called "**Slavic**" **Golden** "**Group**". This group was established in Kiev, the capital of Ukraine, in 1992, during the First International Workshop "**The Golden Proportion and Problems of Systems Harmony**". In 2003, this group was transformed into the *International Club of the Golden Section*. In 2005, according to the initiative of this club, the *Institute of the Golden Section* was organized at the Academy of Trinitarism (Russia). By the way, this institute is the first in the history of science institutes created for the study of the "*golden section*", *Fibonacci numbers* and their applications in modern science. Many outstanding scientists joined the activities of the *International Club of the Golden Section* and the *Institute of the Golden Section*. One of them was the Doctor of Physical and Mathematical sciences, the prominent Russian mathematician, Professor Samuil Aranson (US, San Diego).

Samuil Aranson joined the research in the field of the "*golden section*", *Fibonacci numbers* and "*Mathematics of Harmony*" in 2007. He was immediately interested in the mathematical researches in the "*golden section*", which had a definite relation to his previous professional research, in particular, to mathematical results such as the hyperbolic Fibonacci and Lucas functions [75], Spinadel's metallic proportions [33], Gazale's formulas [34] and so on.

It was these results that led to the "golden" interpretation of the special theory of relativity and the evolution of the Universe starting with the *Big Bang* and ending the original solution of *Hilbert's Fourth Problem* concerning the *hyperbolic geometry* [84–90].

5.3.2. The classical metric form of the Lobachevsky plane

As it is known, the classical metric form of the Lobachevsky plane in pseudo-spherical coordinates (u, v), $0 < u < +\infty$, $0 < v < +\infty$, which has the *Gaussian curvature* $K = -1$, is written as follows:

$$(ds)^2 = (du)^2 + sh^2(u)(dv)^2, \qquad (5.1)$$

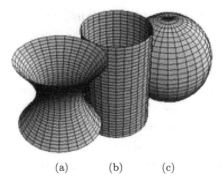

(a)   (b)   (c)

Fig. 5.2. Surfaces with different curvatures: (a) single-cavity hyperboloid, (b) cylinder, (c) sphere.

where ds is the *element of length*, and $\text{sh}(u)$ is a hyperbolic sine. As it follows from (5.1), the *hyperbolic sine* plays a key role in the expression (5.1).

Earlier, we used the concept of the *Gaussian curvature*. What is this? To give the most popular explanation of this rather complex mathematical notion, we consider various surfaces, as shown in Fig. 5.2.

The surface of the *left* figure (*single-cavity hyperboloid*) has a *negative curvature*, the surface of the *center* figure (*cylinder*) has no curvature and is a surface with *zero curvature*, and finally *the surface figure of the right* (*sphere*) is a surface with *positive curvature*. Various approaches have been proposed for measuring the curvature of the surface at a given point. One of them was proposed by Gauss. *Lobachevsky's geometry* leads to the surfaces with *negative curvature*, in particular, with the *Gaussian curvature* $K = -1$.

5.3.3. The metric λ-form of the Lobachevsky plane

It is easy to show that for the general case $\lambda > 0$, the metric λ-form of the Lobachevsky plane is written as follows:

$$(ds)^2 = \ln^2(\Phi_\lambda)(du)^2 + \frac{4 + \lambda^2}{4}[\text{sF}_\lambda(u)]^2(dv)^2, \qquad (5.2)$$

Table 5.1. Metric λ-forms of the Lobachevsky plane.

Title	λ	Φ_λ	Analitical expression
Lobachevsky metric λ-form	$\lambda > 0$	$\Phi_\lambda = \frac{\lambda + \sqrt{4+\lambda^2}}{2}$	$(ds)^2 = \ln^2(\Phi_\lambda)(du)^2$ $+ \frac{4+\lambda^2}{4}[\mathrm{sF}_\lambda(u)]^2(dv)^2$
The "golden" form	$\lambda = 1$	$\Phi_1 = \frac{1+\sqrt{5}}{2} \approx 1.61803$	$(ds)^2 = \ln^2(\Phi_1)(du)^2$ $+ \frac{5}{4}[\mathrm{sF}s(u)]^2(dv)^2$
The "silver" form	$\lambda = 2$	$\Phi_2 = 1 + \sqrt{2} \approx 2.1421$	$(ds)^2 = \ln^2(\Phi_2)(du)^2$ $+ 2[\mathrm{sF}_2(u)]^2(dv)^2$
The "bronze" form	$\lambda = 3$	$\Phi_3 = \frac{3+\sqrt{13}}{2} \approx 3.30278$	$(ds)^2 = \ln^2(\Phi_3)(du)^2$ $+ \frac{13}{4}[\mathrm{sF}_3(u)]^2(dv)^2$
The "copper" form	$\lambda = 4$	$\Phi_4 = 2 + \sqrt{5} \approx 4.23607$	$(ds)^2 = \ln^2(\Phi_4)(du)^2$ $+ 5[\mathrm{sF}_4(u)]^2(dv)^2$
Classical form	$\lambda_e \approx 2.350402$	$\Phi_{\lambda_e} = e \approx 2.7182$	$(ds)^2 = (du)^2$ $+ \mathrm{sh}^2(u)(dv)^2$

where $\Phi_\lambda = \frac{\lambda + \sqrt{4+\lambda^2}}{2}$ represents the *metallic proportions* and $\mathrm{sF}_\lambda(u)$ is the *hyperbolic Fibonacci λ-sine* (4.53). The forms (5.2) are named in Refs. [68–70] as the **metric λ-forms of the Lobachevsky plane**.

Table 5.1 summarizes the expressions for the above-considered metric λ-forms of the Lobachevsky plane for the partial cases $\lambda = 1, 2, 3, 4$.

The overall result of the research, carried out in Refs. [68–70], is the fact that an infinite set of the metric λ-forms of the Lobachevsky plane is obtained; these metric λ-forms ($\lambda > 0$) are given by the general expression (5.2). All these metric λ-forms are isometric to the classical metric form of the Lobachevsky plane given by the expression (5.1). This means that the new models of the Lobachevsky plane [68–70], based on the "metallic proportions" (5.44), together with the classical geometries of Lobachevsky, Riemann and other well-known non-Euclidean geometries *"can be considered as the closest geometries to the ordinary Euclidean geometry*" (David Hilbert).

Thus, the above scientific results, without doubts, are a certain contribution to the solution of *Hilbert's Fourth Problem*, which is one of the *most complicated Hilbert Problems*. It is clear that this solution cannot be treated as the complete solution to this important mathematical problem. Therefore, this solution will undoubtedly stimulate mathematicians to search its new solutions, in particular, the complete solution of *Hilbert's Fourth Problem*.

5.3.4. A summary of the dramatic history of the solution of Hilbert's Fourth Problem in the 20th and 21st centuries

As shown above, the history of the solution of *Hilbert's Fourth Problem* begins from the 1901 PhD thesis defended by Hammel under Hilbert's scientific leadership. However, Hammel's solution did not satisfy the mathematical community and therefore the research in this area was continued, in particular, by the famous Soviet mathematician academician Pogorelov, the American mathematician Buseman, as well as the Ukrainian researcher Stakhov and the Russian mathematician Aranson. The results of this study are given in the Wikipedia paper *"Hilbert's problems"*. In the **Table of problems** of this paper, we find the following conclusion concerning *Hilbert's Fourth Problem*:

"**Brief explanation:** *Construct all metrics where lines are geodesics.*

Status: *Too vague to be stated resolved or not.*"

The "status" of Hilbert's Fourth Problem, given in the above-mentioned table (a *vague problem*), can be interpreted as follows: The mathematicians of the 20th century proved to be incapable of solving Hilbert's Fourth Problem and therefore they decided to pin the whole responsibility for the solution of the Fourth Problem on Hilbert himself because he formulated it *very vague*.

5.4. Complete Solution of Hilbert's Fourth Problem, and New Challenges for the Theoretical Natural Sciences

5.4.1. Insolvability of the Fourth Hilbert Problem for hyperbolic geometries

In 2019, the *Journal of Advances in Mathematics and Computer Science* published a very unusual paper concerning Hilbert's Fourth Problem [151].

This paper proves the insolvability of the Fourth Hilbert Problem for hyperbolic geometries. It has been hypothesized that this fundamental mathematical result (the insolvability of the Fourth Hilbert Problem) holds for other types of non-Euclidean geometry (Riemannian geometry (elliptic geometry), non-Archimedean geometry, and Minkowski geometry). The ancient "golden section", described in Euclidean *Elements* (Proposition II.11) and the following *Mathematics of Harmony* [6], as a new direction in geometry, are the main mathematical apparatus for this fundamental result.

By the way, this solution is reminiscent of the insolvability of the 10th Hilbert Problem for Diophantine equations in integers. This outstanding mathematical result was obtained by the talented Russian mathematician Yuri Matiyasevich in 1970 [150] by using the Fibonacci numbers, introduced in 1202 by the famous Italian mathematician Leonardo from Pisa (by the nickname Fibonacci), and the new theorems in the Fibonacci numbers theory, proved by the outstanding Russian mathematician Nikolay Vorobyev and described by him in one of the latest editions of his book *Fibonacci numbers* [8].

The main result of the paper [151] is obtaining the complete solution of the Fourth Hilbert Problem for the hyperbolic geometries; the essence of this result is formulated as the following theorem.

Theorem. *The Fourth Hilbert Problem is insoluble for hyperbolic geometries.*

The question arises: Does the solution of *Hilbert's Fourth Problem* have some applied significance? Before answering this question, we note that all the formulas of Table 5.1 as well as the formulas of Tables 4.6 and 4.7 and other remarkable identities, given in Chapter 4, evoke a sense of aesthetic pleasure. The mathematical beauty of these formulas is fascinating; this means that these formulas fully satisfy Dirac's *Principle of Mathematical Beauty* [154].

The theoretical significance of these formulas is beyond doubt. It is a striking fact that the number of new hyperbolic λ-functions with irrational bases $\Phi_\lambda = \frac{\lambda+\sqrt{4+\lambda^2}}{2}$ is theoretically infinite because every irrational number $\lambda > 0$ "generates" the new hyperbolic λ-function with irrational base $\Phi_\lambda = \frac{\lambda+\sqrt{4+\lambda^2}}{2}$.

However, the question arises: What hyperbolic λ-functions are the most important from an application point of view? To answer this question, we must once again turn out to the *Bodnar geometry* [28]. **The "Bodnar geometry" shows that the "world of phyllotaxis", one of the most amazing phenomena of botany, is the distinctive "hyperbolic world", based on the hyperbolic Fibonacci and Lucas functions; in this case, the classic "golden proportion" is the base of this "hyperbolic world".** At the same time, this hyperbolic world includes a huge number of real botanical objects observed by us in the wild Nature: *pine and cedar cones, pineapples, cacti, sunflower heads, flower baskets, trees,* etc.

This means that in the botanical phenomenon of phyllotaxis, "hyperbolicity" manifests itself in the *"gold"*, that is, in the *golden proportion* and the *"golden" hyperbolic functions*. This hypothesis, formulated by Oleg Bodnar, proved to be very fruitful and led modern science to the creation of a new geometric theory of phyllotaxis, which has a huge interdisciplinary applied significance [28].

However, the *Fibonacci and Lucas hyperbolic functions* are the partial cases of the *hyperbolic Fibonacci and Lucas λ-functions* (4.53)–(4.56). The latter are based on the *metallic proportions* (4.21), in particular, on the *"silver"*, *"bronze"*, *"copper"* and other types of the *metallic proportions.*

In this regard, we have every reason to suggest that other types of the hyperbolic functions, given by (4.53)–(4.56), can become the basis for modeling of new "hyperbolic worlds", which may really exist in Nature. However, contemporary science until now had not discovered these "hyperbolic worlds" because the contemporary mathematicians did not know about the existence of the new classes of hyperbolic Fibonacci and Lucas functions and no one set such a task for them.

Based on the brilliant success of the "Bodnar geometry" [28], we can set for theoretical physics, chemistry, crystallography, botany, biology, and other branches of the theoretical natural sciences the next challenge of finding new "hyperbolic worlds" of Nature, based on other classes of hyperbolic Fibonacci and Lucas functions, given by (4.53)–(4.56). This is the next important challenge posed by the new theory of hyperbolic λ-functions for modern theoretical natural sciences.

Possibly, the hyperbolic ($\lambda = 2$)-functions (the "silver" functions), which is the second in the list of the hyperbolic λ-functions, is very important from an application point of view.

As mentioned above, the *Bodnar geometry* is based on the *"golden" proportion* and the *"golden" hyperbolic functions*. However, it is possible that the *"golden" proportion* can be the basis of not only the phenomena of phyllotaxis but also other phenomena of Nature or new scientific discoveries. **This is why, for this case, the first challenge to theoretical natural sciences is searching for other phenomena of Nature or scientific discoveries that are based on the *"golden" proportion* and the *"golden" hyperbolic functions*.**

5.4.2. The "silver" proportion as the next challenge for theoretical natural sciences (Tatarenko's proposal)

At the same time, perhaps, the *"silver" proportion* $\Phi_2 = 1 + \sqrt{2}$ and the *"silver" hyperbolic functions* (4.60)–(4.63), based on the *"silver" proportion*, are the next candidates for the "revolution" in the theoretical natural sciences.

The interest to the *"silver" proportion* $\Phi_2 = 1 + \sqrt{2}$ and to the *"silver" hyperbolic functions* have increased significantly in the recent years. In this regard, the article [141] by the Russian researcher Alexander Tatarenko is of particular interest.

In the article [141], Alexander Tatarenko develops the theory of the so-called T_m-*harmonies*, which essentially coincide with Spinadel's *metallic proportions*. At the same time, he assigns a special role of the "*silver*" *proportion* $\Phi_2 = 1 + \sqrt{2}$ (T_2-*harmony*) in further development of the theoretical natural sciences. Alexander Tatarenko claims the following in the article [141]:

> "*The most important and unexpected result of the T_m-harmonies study was the establishment of the two facts:*

> (1) *the second $T_{m=\pm 2} = \sqrt{2} \pm 1$-harmony (not the first, according to numbering in the list of $T_{\pm m}$-harmonies, classical "golden proportion" Φ is the dominant, reigning in the boundless world of the T_m-harmonies.*
> (2) *the "function" of the second $T_{m\pm 2}$-harmony is the number $\sqrt{2}$, the **relic number**, which is found in the archly-vast set of the formulas and laws of the various fields of natural sciences; this is equivalent to the involvement of the T_2-harmony directly or indirectly to the set, and possibly to all the laws of Nature and its constants.*"

Thus, according to Alexander Tatarenko [141], the T_2-harmony literally permeates the entire Universe by its supporting framework and its super fundamental constant, which does not know the limitations, inherent to all the known physical constants without exception.

Alexander Tatarenko expressed his admiration to the T_2-*harmony* in the following remarkable words:

> "*Establishing the fact of the dominance of T_2-harmony and the special status of its "function" $\sqrt{2}$ is the final accord, the most important scientific breakthrough on the path to Truth about the Harmony of the World, comparable to the changing of the Ptolemy geocentrism by the Copernican helio-system. A radically new thinking about the Harmony of the Universe is required.*"

Thus, Tatarenko pays special attention to the *"silver"* proportion $T_2 = 1 + \sqrt{2}$, which *"literally permeates the entire Universe, being its bearing framework and its super fundamental constant, which does not know the limitations, inherent in all the well-known physical constants without exception"*.

Moreover, he considers the implementation of the T_2-*harmony* or the *"silver"* proportion $T_2 = 1 + \sqrt{2}$ and its *"function"* $\sqrt{2}$ into the modern science as *"**the most important scientific breakthrough on the path to Truth about the Harmony of the World, comparable to the changing of the Ptolemy geocentrism by the Copernican helio-system**"*.

5.4.3. The mathematical constants $T_2 = 1 + \sqrt{2}$ and $\sqrt{2}$, Pell numbers and Pythagoras constant

Careful analysis of the mathematical constants $T_2 = 1 + \sqrt{2}$ and $\sqrt{2}$ shows that these constants were in fact widely known in ancient science. The constant $T_2 = 1 + \sqrt{2}$, named as the *"silver"* proportion in Spinadel's book [33], and the T_2-*harmony* in Tatarenko's paper [141], go back in their origins to the so-called *Pell numbers*, which are described by many famous researchers in this field. The constant $\sqrt{2}$ was well known in ancient science under the title *Pythagoras's constant*.

We can read in the Wikipedia paper "Pell number" the following interesting information about this numerical sequence:

> *"In mathematics, the **Pell numbers** are an infinite sequence of integers, known since ancient times, that comprise the denominators of the closest rational approximations to the square root of 2. This sequence of approximations begins $\frac{1}{1}$, $\frac{3}{2}$, $\frac{7}{5}$, $\frac{17}{12}$, and $\frac{41}{29}$, ... so the sequence of Pell numbers begins with 1, 2, 5, 12, and 29. The numerators of the same sequence of approximations are half the **companion Pell numbers** or **Pell–Lucas numbers**; these numbers form a second infinite sequence that begins with 2, 6, 14, 34, and 82.*
>
> *Both the Pell numbers and the companion Pell numbers may be calculated by means of a recurrence relation, similar to that for the Fibonacci numbers, and both sequences of numbers grow*

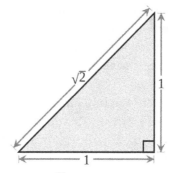

Fig. 5.3. Pythagoras' constant $\sqrt{2}$. (the square root of $2(\sqrt{2})$ is equal to the length of the hypotenuse of a right-angled triangle with the legs of length 1).

exponentially, proportionally to powers of the silver ratio $1 + \sqrt{2}$. As well as being used to approximate the square root of two, Pell numbers can be used to find square triangular numbers, to construct integer approximations to the right isosceles triangle, and to solve certain combinatorial enumeration problems."

The square root of 2, or **Pythagoras constant**, denoted as $\sqrt{2}$, is the positive algebraic number, which, when multiplied by itself $(\sqrt{2} \cdot \sqrt{2})$, gives number 2. Geometrically, the square root of $2(\sqrt{2})$ is the length of a diagonal across a square with the sides of one unit of length; this follows from the Pythagorean Theorem. (see Fig. 5.3).

It is important to note that *Pythagoras constant* $\sqrt{2}$ is mentioned in the Wikipedia paper "Mathematical constant" in one row with such important mathematical constants as *Archimedes's constant* π, *Euler's number e*, *the golden ratio* and other famous constants of the advanced mathematics.

Therefore, we must relate with proper respect to the above-considered quote by Tatarenko, in which he compares the implementation of the mathematical constants $T_2 = 1 + \sqrt{2}$ and $\sqrt{2}$ into modern science *"with the most important scientific breakthrough on the path to Truth about the Harmony of the World, comparable with the replacement of the Ptolemy geo-centrism by the Copernicus helio-system."*

5.5. New Approach to the Creation of New Hyperbolic Geometries: From the "Game of Postulates" to the "Game of Functions"

According to Stakhov's article [68], the cause of the difficulties, arising at the solution of *Hilbert's Fourth Problem*, lies elsewhere. All the most known attempts to resolve this problem (Herbert Hamel and Alexey Pogorelov) were based on the traditional approach, which reduces to the so-called *game of postulates* [68]. The *game of postulates* in geometry began from the works of Nikolay Lobachevsky and Janos Bolyai, when the *classical Euclid's postulate of the parallel lines* was replaced with the opposite one. This was the most major step in the development of the *non-Euclidean geometry*, which led to *Lobachevsky's geometry*. According to the outstanding Russian mathematician Kolmogorov [102], *Lobachevsky's geometry* is considered as the **most important mathematical discovery of the 19th century** and can rightly be compared to the **MILLENNIUM PROBLEM in geometry [96]**. It changed our traditional geometric ideas about geometry and led to the creation of the *hyperbolic geometry*. It must be emphasized that the title of *hyperbolic geometry* highlights the fact that this geometry is based on the *hyperbolic functions*. **The use of the *hyperbolic functions* for the mathematical description of *Lobachevsky's geometry* became one of the "key" ideas in the creation of the *hyperbolic geometry.***

Chapter 6

Beauty and Aesthetics of Harmony Mathematics

6.1. Mathematics: A Loss of Certainty and Authority of Nature

6.1.1. Morris Klein's book

In 1980, Oxford University Press (New York) published the book *Mathematics. The Loss of Certainty*, written by the famous American historian of mathematics Morris Klein (1908–1992), the Emeritus Professor of Mathematics of the Courant Institute of Mathematical Sciences of New York University. In 1984, this book was translated into Russian [101]. Morris Klein's book forced modern mathematicians to think deeply about the deplorable state of modern mathematics, in which it found itself after the emergence of the new crisis in its foundations, which arose in the early 20th century in connection with the discovery of paradoxes in *Cantor's theory of infinite sets*. Unfortunately, this crisis has not been overcome till the present time. Morris Klein wrote the following about this:

> *"At the moment, the state of affairs in mathematics can be described as follows. There exist not one, but many mathematics, and each of them, for different reasons, does not satisfy mathematicians, belonging to other schools. It became clear that the idea about the vault of generally accepted, unshakable truths, that is, about the majestic mathematics of the early 19th century, the pride of mankind, is nothing more than a delusion. Instead confidence and wellbeing that prevailed in the past, there had come*

the insecurity and doubts about the future of mathematics come. Disagreements over the foundations of the most 'unshakable' of the sciences caused surprise and disappointment.... **The current state of mathematics is nothing more than a pathetic parody on the mathematics of the past with its deep-rooted and well-known reputation as an impeccable ideal of truth and logical perfection.***"*

6.1.2. The Authority of Nature

There arose the following question: How can one overcome the crisis in modern mathematics? Morris Klein answers this question in the final chapter "The Authority of Nature" of his book [101] as follows. For the development of certain mathematical directions, mathematicians are forced to *be guided by external considerations.* At the same time, the significance of mathematics for other sciences is "*the traditional and most explicable arguments in favor of creating new and developing already existing mathematical theories.*"

Klein emphasizes as follows [101]:

"The applications serve as a kind of practical criterion by which we test mathematics... Why not judge now about the correctness of mathematics in general by how well it continues to describe and predict natural phenomena?"

Morris Klein quotes in the book [101] the following statement of the British philosopher, political economist, and civil servant John Stuart Mil (1806–1873):

"Those who think that mathematical theorems are qualitatively different from the hypotheses and theories of other areas of science are deeply mistaken."

The same ideas were advocated by one of the outstanding specialists in the foundations of mathematics, Polish mathematician Andrzej Mostowski (1913–1975). He argued that:

"mathematics is connected deeply with Nature. Its basic concepts and methods go back to experience, and any attempt to substantiate mathematics without regard to its

Natural scientific background, applications, or even its history is doomed to failure" [101].

Hermann Weil (1885–1955) adhered the same point of view; according to Morris Klein, he *"openly advocated considering mathematics as one of the natural sciences"* [101].

In the well-known article "The Mathematician", John von Neumann tried to explain why most mathematicians continue to use classical mathematics [101]:

> *"In the end, it is classical mathematics that allows to get results that are both useful and beautiful, and although is not now previous confidence in its reliability, classical mathematics rests on an equally strong basis, like to the existence of an electron. Consequently, the somebody, who accepts the natural sciences, also accepting the classical system of mathematics."*

Morris Klein concludes thus [101]:

> *"The result of all this turbulent and diverse activity was the conclusion that true mathematics should not be determined by grounds, the accuracy of which can be challenged; the "correctness" of mathematics should be judged by its applicability to the real world. Mathematics is the same empirical science as Newtonian mechanics. Mathematics is correct, until then, when it works, and if something does not work, then it is necessary to introduce the appropriate amendments. Mathematics is not a compilation of a priori knowledge, which was believed to be unchanged for more than two thousand years; the mathematics is not absolute and not unchanged."*

6.1.3. Appeal to the origins of mathematics

Overcoming the crisis in modern mathematics requires turning to the origins of mathematics. According to the prominent Soviet mathematician, Andrey Kolmogorov [102],

> *"a clear understanding of the independent state of mathematics as a special science, having its own subject and method, became possible*

only after accumulating a sufficiently large amount of the factual material; this situation first aroused in the Ancient Greece in the 6th and 5th centuries BC. The development of mathematics before this time should naturally be attributed to the period of the origin of mathematics, and the period between the 6th and 5th centuries BC we should consider as the beginning of the period of elementary mathematics."

Morris Klein claims as follows in the book [101]:

*"**The real goal of the ancient Greeks was the study of Nature. This goal was served by everyone: even geometric truths were highly valued only in so far as they were useful in studying the physical world.** The Greeks understood that geometrical principles are embodied in the structure of the Universe, the primary component of which is space. That is why the study of space and spatial figures was a significant contribution to the study of Nature. Geometry was an integral part of the broader program of cosmological research. Similar facts and a more complete knowledge of how mathematics developed in subsequent times allowed us to state that the Greeks had natural-science studies to formulate mathematical problems and that mathematics was the integral part of nature studies."*

In the 2011 book by Dimitrov [44], these ideas of Morris Klein are defined concretely as follows:

*"Harmony was the key concept of the Greeks, with which the three meanings were connected. Its root value was **aro**, a compound; harmony was something, what unites. Another meaning was the proportion, the balance of things, which allowed a simple connection. The quality of the connection and the proportions later began to be seen in music and other forms of art.*

The prerequisite for harmony for the Greeks was expressed in the phrase "nothing superfluous". This phrase contained mysterious positive qualities that have become the object of the study of the best minds. Thinkers, such as Pythagoras, sought to uncover the mystery of harmony as something ineffable and illuminated by mathematics. The mathematics of harmony, studied by the ancient Greeks, is still an inspiring model for modern scholars.

Crucial to this was the discovery of quantitative expression of harmony, in all the amazing diversity and complexity of nature, through the golden ratio $\Phi = \frac{1+\sqrt{5}}{2}$, what is approximately equal to 1,618."

Thus, in the center of the mathematical doctrine of Nature, created by the ancient Greeks, was the *concept of Harmony*, and the mathematics of the ancient Greeks itself was *the Mathematics of Harmony*, which was directly connected with the *golden section*, the most important mathematical discovery of ancient science in the field of *Harmony*. If we want to build a new mathematics, devoid of contradictions, we must decisively introduce into mathematics Pythagorean "Mathem of Harmonics" based on the *Idea of Harmony* and the *golden ratio*.

6.2. Strategic Mistakes in the Development of Mathematics: The View from the Outside

6.2.1. The moving away of mathematics from theoretical natural sciences

What is mathematics? What is its origin and history? What is the relationship and difference between mathematics and other scientific disciplines? All these questions have always interested both mathematicians and representatives of other sciences. Mathematics has been considered always as an example of scientific rigor. It was often called the *Queen of Sciences*, thus emphasizing its special status in science. This is why the emergence of the book *Mathematics. The Loss of Certainty* [101], mentioned above, proved to be a real shock for mathematicians. This section is devoted to the analysis of the *global crisis*, into which mathematics landed in the 20th century as a result of its *illogical development*.

Morris Klein claims as follows in this book [101]:

"The history of mathematics knows not only the greatest flights up, but also deep falls down. The realizing of the fact that the showcase of the human mind, which is sparkling by its splendour, is very far from perfection by its structure, suffers from many flaws and is

subjected to monstrous contradictions, brought one more blow on status of mathematics.

But the disasters that crashed on mathematics were caused by other reasons. Severe presentiments and disagreements between mathematicians were made conditional due to the very course of the development of mathematics in the last hundred years. Most mathematicians as it were fenced oneself from the outside world, by concentrating their efforts on the problems that arise within mathematics itself; in essence, they broke with the natural sciences."

Further, he states thus:

*"Natural sciences were the blood and flesh of mathematics and nourished it with life-giving juices. Mathematicians willingly collaborated with physicists, astronomers, chemists and engineers in solving various scientific and technical problems, and often themselves were outstanding physicists and astronomers. In the 17–18th centuries and also for the most part of the 19th century, the distinction between mathematics and theoretical natural sciences was extremely rare. Many leading mathematicians, by working in the field of astronomy, mechanics, hydrodynamics, electromagnetism and the theory of elasticity, obtained here incomparably more important results than actually in mathematics. **Mathematics was a queen and at the same time a servant of the natural sciences.**"*

Klein emphasizes that the problems of the *pure mathematics*, which came to the fore in the 20th century, did not attract much interest from our great predecessors. Klein cites the opinion of the Francis Bacon (1561–1626) about the *pure mathematics*:

*"The criticism of the 'pure mathematics', that is, the mathematics for the sake of mathematics, can be found in the work of Francis Bacon '**On the Dignity and Multiplication of Sciences**' (1620). Bacon objected against pure, mystical and self-satisfied mathematics, 'completely abstracted from matter and physical axioms,' by complaining on that fact that **there is such property of the human mind: without sufficient efforts to solving important problems, it spends himself on the nonsense.**"*

In 1895, Felix Klein (1849–1925), who was at that time the recognized leader of the mathematical world, also found it necessary to protest against abstract, pure mathematics [51]:

"It is difficult to separate from the feeling that the rapid development of modern mathematics carries with it the danger of increasing of isolation for our science. The close relationship between mathematics and theoretical natural sciences, which existed for the greatest benefit for both sides, threatens to be interrupted due the emergence of modern analysis."

Richard Courant (1888–1972), who headed the Institute of Mathematical Sciences at New York University, also disapproved of the fascination with pure mathematics. In 1939, he wrote thus [51]:

*"A serious threat to the very life of science stems from the statement that mathematics is nothing but a system of conclusions, derived from definitions and postulates, which must be non-contradictory, and otherwise to be arbitrary outcomes of free will of mathematicians. If such a description would true, then in the eyes of any reasonable person, mathematics would not have any attractiveness. It would be an unprovoked aimless play with definitions, rules, and syllogisms. The notion that the mind can arbitrarily create sensible axiomatic systems at its own discretion is a half-truth that can only introduce inexperienced people to a delusion. **Only a free mind held back by the discipline of responsibility to an organic whole, guided by inner necessity, can create results of scientific significance.**"*

Thus, following Bacon, Klein, Courant and other outstanding scientists and thinkers, Morris Klein saw the reason for the current crisis in mathematics in its breaking from natural sciences, which for centuries were the main source for the development of mathematics. The breaking of mathematics from theoretical natural sciences is the largest "strategic error" of mathematics in the 20th century and the main cause of its modern crisis. But in the history of mathematics, there were also other "strategic mistakes", the analysis of which is given in the following sections.

6.2.2. Neglect of "beginnings"

Andrey Kolmogorov, in the preface to Lebeg's book *On the Measurement of Magnitudes* [113], notes that

> *"mathematicians have a tendency, by having already mastered a complete mathematical theory, to be ashamed of its origin. Compared to the crystalline clarity of the theory, by starting from its basic concepts and assumptions, it seems like a dirty and unpleasant affair to delve into the origin of these basic concepts and assumptions.*
>
> *The whole building of school algebra and all mathematical analysis can be created on the concept of real number without any mention of the measurement of specific magnitudes (lengths, areas, time intervals, etc.). Therefore, at different stages of learning with different degrees of courage, the same tendency is manifested: as possible as faster there is to introduce a conception of numbers and further to talk only about the numbers and the ratios between them. Lebeg protests against this tendency."*

In this statement, Kolmogorov noticed one feature of mathematicians: a bashful attitude toward the "beginnings" of mathematical science, or rather *neglecting of "beginnings"* (*at different stages of learning and with varying degrees of courage*). Long before Kolmogorov, Nikolay Lobachevsky drew attention to this peculiarity of mathematicians. In one of his works, he wrote:

> *"Algebra and Geometry have one and the same fate. After early successes, there followed very slow ones, which left science at the stage, where it is still far from perfect. This was due to the fact that the Mathematicians turned their attention to the highest parts of Analysis, by neglecting the beginnings and not wishing to work on the processing of such a field, which they had already gone over and left behind."*

But it was Lobachevsky who, through his research, showed that the "beginnings" of mathematical science, in particular, the

Euclidean *Elements*, are an inexhaustible source of new mathematical ideas and discoveries. Lobachevsky begins his famous *Geometric Studies on the Theory of Parallel Lines* (1840) with the following words:

> *"In geometry, I found some imperfections, which I consider as the reason that this science has so far not gone one step beyond the state, in which it passed to us from Euclid. I attribute to these imperfections the vagueness in the first concepts about geometric magnitudes, the ways, in which we imagine the measurement of these magnitudes, and, finally, an important gap in the theory of parallel lines. . ."*

As it is known, Lobachevsky, unlike other mathematicians, did not neglect the "beginnings". A careful study of the Fifth Euclidean postulate (*an important gap in the theory of parallel lines*) led Lobachevsky to creation of the *non-Euclidean geometry*, which, according to academician Kolmogorov [52], is considered as the largest mathematical achievement of the 19th century.

6.2.3. Neglect of the "idea of harmony" and the "golden section"

As mentioned in Vol. I, a special role in the Pythagoras and Plato Doctrines is played by the *golden section*, which at that time was called the *division of a segment by the extreme and mean ratio*. The "golden section" literally permeates the Euclidean *Elements*, starting from Book II (*Proposition II.11, the golden section*) and ending with Book XIII (*Platonic solids*).

As mentioned in Vol. I, the *golden section* played a *key* role in the Pythagoras and Plato Doctrines; the "golden section" at that time was called as the *task of the division of a segment in the extreme and mean ratio* [32]. This task literally permeates the Euclidean *Elements*, starting from Book II (Proposition II.11) and ending with Book XIII, devoted to the geometric theory of *Platonic solids*.

6.2.4. The golden ratio in natural science

Neglect of the *golden section* was found not only in mathematics but also in theoretical physics. In 2006, BIMON (Moscow) published a collection of scientific articles *Metaphysics. Century XXI* [142]. The book [142] consists of three parts:

Part I. General issues of metaphysics;
Part II. Metaphysics and research programs in theoretical physics;
Part III. The golden ratio in natural science.

In the preface to the book [142], the compiler and editor of this collection, the famous Russian scientist and thinker Professor Yury Vladimirov (Moscow University) wrote the following:

> *"The third part of the collection is devoted to the discussion of the many examples of the manifestation of the golden ratio in art, biology and in the reality around us. However, paradoxically, in modern theoretical physics the golden ratio isn't reflected in any way. To verify in this, it is enough to scroll through the 10-volume book of theoretical physics by Landau and Lifshitz. The time is ripe to fill this gap in physics, especially because the golden ratio is closely related to metaphysics and Trinitarism.*
>
> *... **Stakhov's article "The Golden Ratio, Sacred Geometry and Mathematics of Harmony", which summarizes the history of the "golden ratio", points its most important manifestations and attitudes to it over the centuries, opens this part of the article collection [142].***
>
> *The Golden Section and the associated to it Fibonacci numbers permeate the entire history of art. The Cheops Pyramid, the most famous of the Egyptian pyramids, the famous Greek temple of Parthenon, the majority of sculptural monuments, the unsurpassed "Mona Lisa" by Leonardo da Vinci, the paintings by Raphael and Shishkin — these are far not a complete list of the outstanding works of art, filled with wonderful harmony based on the "golden section".*
>
> *The article sets forth the foundations of a new theory of hyperbolic Fibonacci and Lucas functions, which are built on the basis of formulas corresponding to the well-known formulas of trigonometry and classical hyperbolic functions. In them, instead*

of the number e — the base of the natural logarithms — the value of the "golden section" is used Φ... It is shown that, on the basis of a new class of hyperbolic functions, a number of patterns in the evolution of plants and animals, in particular, phyllotaxis phenomena, are naturally described. Other applications of this theory in computer technology, in the tasks of information coding, digital signal processing, etc. is shown."

A very interesting explanation of the reason about the scornful (in some sense even hostile) attitude of official academic science toward *harmony* and *golden section* is contained in the article by Denis Kleschev [115]. Kleschev writes thus:

"The unconditional success of Mendeleev's theory, which modern priests of science stubbornly retouch, is explained by the fact that the 'speculative' splitting of the periodic table into seven periods reflects the objective existence in the atomic structure of seven energy levels ('shells'). These energy levels together with the core form a harmonic 'octave', filled with electrons. The transition of an electron from one of the outer levels to a closer to the nucleus is always associated with the emission of a quantum of light; therefore the language of the spectrum was and remains one of the most important tools of nuclear physics."

The effectiveness of introducing the *quantum numbers* so much impressed the pioneers of atomic physics that the great German theoretical physicist Arnold Sommerfeld (1868–1951) wrote the following in the first edition of his fundamental work *Atomic Structure and Spectral Lines* (1919):

"What we hear today in the language of spectra is genuine music of the spheres of atoms, the consonance of integer relations, one of the many manifestations of ever-increasing order and harmony.

All the integer regularities of spectral lines and atomistic originate, ultimately, from quantum theory. It is the mysterious organ, on which Nature plays the music of the spectra, and the structure of the atom and nucleus obeys to this rhythm."

In one of his speeches (1925), Arnold Sommerfeld exclaimed, not without bitterness:

"If only Kepler could live till modern quantum theory! He could see how came true his wildest youthful dream, but not in the macrocosm of celestial solids, but in the microcosm of the atom."

We continue to the following quote of Denis Kleschev [115]:

"However, such interpretations of the atomic theory were heavily criticized. The most prominent astrophysicist Arthur Eddington (since 1923 foreign correspondent member of the USSR Academy of Sciences, President of the International Astronomical Union since 1938, suffered from the spread of 'Pythagoreanism' during the formation of modern normal science. His attempts to link the fine structure constant $\alpha = 1/137\,035$ with the 'theory of harmony' initiated a loud process in the USSR Academy of Sciences; as a result any mention of the word 'harmony' in quantum physics was forbidden and was automatically understood as 'Eddingtonism'".

Thus, already in the Soviet times, the higher academic circles had a hand in denouncing the concept of "harmony" and removal of this word from physical science. **So not only "cybernetics" and "genetics", but also the "doctrine of harmony" were subjected to persecution by the Soviet official academic science.**

Despite the fact that academic science has opposed to the usage of the concept of "harmony" in modern physics, the development of the natural sciences in the recent decades has shown that respected academicians made mistakes in assessing the role of "harmony" in modern physics.

Since academic science at the initial stage, as a rule, makes mistakes in the evaluation of certain areas of fundamental sciences (recall *genetics* and *cybernetics*, and earlier *Lobachevsky's hyperbolic geometry*) and then usually forgets to repent about what it has done, disrespect by the "harmony" and the "golden section" is another "strategic mistake" in the development of not only mathematics but also theoretical physics. This mistake gave rise to a number of other "strategic mistakes" in the development of mathematics and science in general.

6.2.5. Disrespect to the "Proclus hypothesis"

In Vol. 1, we analyzed in detail the new look at the Euclidean *Elements* put forward by the Greek philosopher Proclus Diadoch (412–485).

The analysis of the *Proclus hypothesis* is contained in many Western books on the history of mathematics. Consider some of them: [117, 121, 122]. The book [117] claims thus:

> *"According to Proclus, the main purpose of Euclid's Elements was to set out the geometric theory of the so-called Platonic solids."*

In the book [121], this idea gets further concretization:

> *"Proclus, by mentioning once again all the previous mathematicians of Plato's circle, says: "Euclid lived later than the mathematicians of Plato's circle, but earlier than Eratosthenes and Archimedes. Euclid belonged to Plato's school and was well acquainted with Plato's philosophy; that is why, Euclid set as the main goal of his Elements to create the geometric theory of the so-called Platonic solids."*

This comment is important for us because it draws attention to the connection of Euclid with Plato. Euclid fully shared *Plato's philosophy* and *Plato's cosmology* based on the *Platonic solids*; this is why **Euclid set the creation of the geometric theory of the Platonic solids** as the main goal of his *Elements*.

In Vol. I, we described the so-called *Cosmic Cup*, the original model of the Solar system, based on the *Platonic solids*. This model was described by Kepler in his first book, *Mysterium Cosmographicum*. We are interested in the *Cosmic Cup* primarily from the point of view of Kepler's relation to the *Proclus hypothesis*. Craig Smorinsky in the book [122] discusses the influence of Plato and Euclid's ideas on Johannes Kepler:

> *"Kepler's project in 'Mysterium Cosmographicum' was giving 'true and perfect reasons for the numbers, values, and periodic motions of celestial orbits'. The perfect reasons should be based on the simple principles of mathematical order, which Kepler found in the Solar system, by using numerous geometric demonstrations. The general scheme of his model was taken by Kepler from Plato's 'Timey', but the mathematical relations for the Platonic solids (pyramid, cube, octahedron, dodecahedron, icosahedron) were taken by Kepler from the works of Euclid and Ptolemy. At the same time, Kepler followed Proclus in the following:* **'the main goal of Euclid was to create a geometric theory of the so-called Platonic solids.'** *Kepler was completely fascinated by Proclus, he often quoted Proclus and calls him the 'Pythagorean'."*

In this connection, it is also appropriate to recall the book *Lectures About the Icosahedron and Solving Fifth-Degree Equations* [113], published in 1884 by the outstanding mathematician Felix Klein (1849–1925). In this book [113], the icosahedron, based on the *golden section*, is considered as the *main Platonic solid*, which plays one of the central roles in mathematics.

Some Soviet mathematicians also expressed ideas close to the *Proclus hypothesis*. In Ref. [128], Denis Kleshchev cites the opinion of Prof. D.D. Mordukhai-Boltovsky (the most authoritative Soviet historian of mathematics and the translator of the Euclidean *Elements* into Russian); Mordukhai-Boltovsky's opinion fully coincides with the *Proclus hypothesis*:

> **"A thorough analysis of Euclid's Elements convinced me that the construction of regular polyhedra, and even more, the proof of the existence of five and only five Platonic solids, once, even before Euclid, was the ultimate goal of the works, from which the Elements originated."**

The question arises: Why is the *Proclus hypothesis* and the opinion of the prominent Soviet historian of mathematics not widely reflected in the historical literature and textbooks in mathematics?

Let's once again turn to the article by Denis Kleshchev [128]. Kleschev writes thus:

"With a careful study of the Euclidean Elements, we can find the harmonic component of this treatise, which is fundamentally different from other treatises on geometry (for example, from the Elements of Hippocrates of Chios). Euclid's treatise earned universal recognition in antiquity by the fact that all thirteen books, in which Euclid gave a systematic presentation of the most significant discoveries of ancient mathematics, had the main purpose finding and proving the existence of the five regular polyhedra (Platonic solids: tetrahedron, hexahedron, octahedron, icosahedron, dodecahedron).

Despite the fact that Euclidean geometry became the basis of classical mathematics and was erected by subsequent generations of mathematicians to the rank of 'absolutely true science', the theory of harmony remained a metaphysical appendage, for which there was no decent place in the scientific paradigm; therefore **today the concept of 'harmony' is associated only with the musical doctrine of the Pythagoreans, although in reality, in ancient science, the harmony of numerical relationships extended to all areas of knowledge.**

The main reason, why the general theory of harmony did not develop within the framework of the ancient paradigm, was the fact, that the study of harmonic relations inevitably led mathematicians to incommensurable segments and irrational numbers, to which ancient philosophers had unconcealed disgust.

It seems impression that modern historians of mathematics hold the same point of view. They refuse to notice that Euclid in his Elements paid great attention to the 'golden section' and 'Platonic solids' and they cannot give any 'scientific explanation' to this fact. **Moreover, they refused to admit the fact that, according to the 'Proclus hypothesis', Euclidean Elements were written under the direct influence of the ancient idea of harmony. And the main goal of Euclid when writing his brilliant mathematical work, which is the foundation of all modern mathematics, was creating a complete theory of Platonic solids, which are associated in the ancient Greeks with the harmony of the Universe...**

Although the 'Proclus Hypothesis' (V century AD), which reveals the meaning of the Euclidean 'Elements', compiled as a strictly axiomatic construction of the five regular polyhedra, has many confirmations in ancient science, modern mathematicians continue to hide this historical fact from us."

In this connection, it is appropriate to quote another statement by Denis Kleschev [128]:

"The theory of harmonization, systematization and generalization of information seems for many mathematicians to be very insignificant problem. Most modern mathematicians consider it extremely boring to study elementary geometry, although none of the outstanding mathematicians of the XIX century (including universal mathematicians Hilbert and Poincaré), did not feel such indifference and disdain to elementary mathematics, to which today feel an average associate professor, a graduate student or a mathematics student. So we are reaping the fruits of the style of teaching that emerged after the introduction of the set-theoretical approach into mathematical education."

Thus, we can state that Proclus's view on Euclidean Elements was ignored by official academic mathematics; moreover, modern Russian historians of mathematics hide this historical fact from the scientific community, which was recognized by Johannes Kepler, Felix Klein, and in our times, Mordukhai-Boltovsky. All this should be considered as another "strategic mistake" in the development of mathematics, which led to a distorted view of the entire history of mathematics.

6.2.6. One-sided look at the origin of mathematics

As it is well known, the traditional view of the origin of mathematics [102] consists in the fact that the creation of mathematics in the ancient period was stimulated by two *"key"* *practical problems* that arose in science at the early stages of its development: the *problem of counting* and the *problem of measurement*. The problem of counting led to the creation of the first methods of representing numbers and performing simple arithmetical operations over numbers (the Babylonian positional numeral system with base 60, Egyptian non-positional decimal arithmetic, etc.). The main result of this process was the formation of the concept of *natural numbers*, the fundamental concept of mathematics, without which the existence of mathematics is unthinkable. The problem of measurement lies at the foundation of geometry as a science about *measuring the Earth*. It was within the framework of the study of the *problem of measurement* in the scientific school of Pythagoras that the incommensurable segments were discovered; this discovery is considered as one of the most important mathematical achievements of ancient mathematics. This discovery led to the introduction of the concept of *irrational numbers*, the second (after natural numbers) fundamental concept of mathematics, without which it is impossible to imagine the existence of modern mathematics.

The concepts of *natural numbers* and *irrational numbers* underlie *Classical Mathematics* and *Classical Natural Sciences*. Note that the most important "mathematical constants", in particular, *the number* π and the *Euler's number* e are irrational (transcendental) numbers. These numbers underlie the most important types of *elementary functions*, in particular, the *trigonometric functions* (the number π) and the *hyperbolic functions* (the Euler's number e).

Unfortunately, the historians of mathematics sometimes forget to mention that there was another scientific problem, the problem of *Harmony*, which was formulated by Pythagoras and Plato and was reflected in the Euclidean *Elements*, according to the *Proclus hypothesis*. The most surprising thing is that even such an outstanding mathematician and historian of mathematics as Andrey Kolmogorov, in his remarkable book [102] doesn't mention the *Proclus hypothesis* and the influence of the *Harmony problem* on the creation of the Euclidean *Elements*.

It is difficult to assume that Academician Kolmogorov didn't know about the *Proclus hypothesis*. Most likely, he simply didn't dare to mention it because in this case, the history of mathematics, set forth in his book [102], would have to be rewritten in a new way.

As a result, in modern mathematics, going back to the Euclidean *Elements* in its origin, the distorted view on the Euclidean *Elements* and the origin of mathematics was formed; this fact is another "strategic mistake" in the development of mathematics history. If the problem of harmony was included in the list of the most important problems that historically influenced on the creation of mathematics at the initial stage of its development, the structure and the content of modern mathematics would look otherwise.

6.2.7. The greatest mathematical mystification of the 19th century

Major "strategic mistakes" were made in subsequent periods of mathematics development. In the 19th century, one of these mistakes was the decision about the inclusion of the *theory of Cantor's infinite sets* to the list of the "greatest mathematical discoveries" without sufficient critical analysis of this theory.

Cantor's theory of infinite sets led to a storm of protests during the early 19th century. A detailed analysis of the criticism of this theory was given in the chapter *"Expulsion from Paradise: A New Crisis in the Foundations of Mathematics"* of the book by Morris Klein [101].

Many famous mathematicians of the 19th century spoke out sharply negatively about this theory. Leonid Kronecker (1823–1891), who had a personal dislike to Cantor, called him a *charlatan*. Henri Poincare (1854–1912) called Cantor's theory of sets a *serious illness* and considered this theory as a kind of the *mathematical pathology*. In 1908, he declared: *"Future generations will see Cantor's set theory as a disease, from which they should been cured."*

Unfortunately, Cantor's theory of infinite sets had not only opponents but also supporters. The great British philosopher, logician, mathematician, historian, writer, essayist, social critic, political activist, and Nobel laureate Bertrand Russell (1872–1970) called Cantor as one of the greatest thinkers of the 19th century. In 1910, Russell wrote:

"Solving problems, which have long enveloped by mystery of mathematical infinity, is probably the greatest achievement that our age should be proud of."

In his speech at the *First International Congress of Mathematicians* in Zurich (1897), the famous French mathematician Jacques Hadamard (1865–1963) emphasized that the main attractive feature of Cantor set theory is the fact that for the first time in mathematical history, the classification of sets based on the concept of *cardinal number* is given. The amazing mathematical results that follow from Cantor's set theory inspire mathematicians on new discoveries.

In the recent years, in the works of the outstanding Russian mathematician and philosopher Alexander Zenkin [126], as well as in the works of other authors [127–129], radical attempts have been made to "purify" mathematics from the Cantor theory of infinite sets. Analysis of the Cantor theory of infinite sets, presented in [126, 127], led Alexander Zenkin to conclusion that the proofs of many Cantor's theorems on infinite sets are logically incorrect, and the whole "Cantor theory" is in some sense *the greatest mathematical mystification of the 19th century.*

The discovery of paradoxes in Cantor's theory of infinite sets considerably cooled the enthusiasm of mathematicians in this theory, but Alexander Zenkin put possibly the final point in the critical analysis of Cantor's theory [126, 127]. He showed that the main Cantor's mistake was the adoption of the abstraction of *actual infinity* which, starting with Aristotle, is unacceptable in mathematics.

But without the abstraction of actual infinity, the theory of Cantor's infinite sets is untenable! Aristotle, who first warned about the impossibility of using the concept of "actual infinity" in mathematics (*"Infinitum Actu Non Datur"*) drew attention to this problem.

Thus, Cantor's theory of infinite sets is nothing more than the greatest *mathematical mystification* of the 19th century, a kind of *mathematical quackery*, and its adoption by the mathematicians of the 19th century, without proper critical analysis, is another "strategic mistake" in the development of mathematics. If Cantor's theory of infinite sets was seriously analyzed in the 19th century, if mathematicians listened authentically to the opinion of the eminent mathematicians Kronecker and Poincaré, it would have been possible to avoid the occurrence of the modern crisis in the foundations of mathematics. Modern historical mathematicians must not ignore the famous saying of Aristotle *"Infinitum Actu Non Datur"*.

6.2.8. Underestimation of Binet formulas

In the 19th century, an important mathematical discovery was made in the development of the *golden section* theory. We are talking about the so-called *Binet formulas*, derived by the French mathematician Binet (1786–1856) in the 19th century. It is surprising that, in Classical Mathematics, Binet's formulas did not receive proper recognition, similar to other well-known mathematical formulas ("Euler formulas", "Moivre formulas", etc.), and not all mathematicians knew about *Binet's formulas*. The study of *Binet's formulas*, as well as the *golden section, Fibonacci numbers* and *Lucas numbers*, as a rule, is not included in modern mathematical programs of high school and university education. Apparently, this attitude to the *Binet formulas* is associated with the *golden section*, which always caused an "allergy" in some mathematicians.

But the main "strategic mistake" in the evaluation of *Binet's formulas* is the fact that mathematicians did not see in these formulas a prototype of a new class of hyperbolic functions, the *hyperbolic Fibonacci and Lucas functions*, which have been discovered by the Ukrainian researchers Bodnar, Stakhov, Tkachenko and Rozin 100 years after the discovery of *Binet's formulas* [28, 64, 75]. If the Fibonacci and Lucas hyperbolic functions had been discovered in the 19th century, then the hyperbolic geometry and its applications in physical science would look different and maybe already in the 19th century a new geometric theory of phyllotaxis was created; in modern science, a new geometric theory of phyllotaxis is called "Bodnar geometry" [28].

6.2.9. Underestimation of the "icosahedral" idea of Felix Klein

As mentioned in Vol. I, in the 19th century, the eminent mathematician Felix Klein tried to unite all branches of mathematics on the basis of *icosahedron*, the *Platonic solid*, dual to the *dodecahedron* [113]. In essence, Klein's research can be viewed as further development of the so-called *dodecahedral–icosahedral idea*, which, starting with Pythagoras, Plato, Euclid and Kepler, penetrates through the entire history of science. Klein treats the *icosahedron*, based on the *golden section*, as a geometric object, from which, in his opinion, the branches of the five mathematical theories follow: *geometry, Galois theory, group theory, invariant theory, and differential equations*. The main idea of Klein is as follows:

> *"Each unique geometric object is somehow connected with the properties of the icosahedron."*

Unfortunately, this deep geometric idea hasn't seen further development in modern mathematics, which is another *strategic mistake* in mathematics development. The development of this idea could affect the structure of mathematical science and could lead to the unification of many important branches of mathematics on the basis of *icosahedron*, connected to the *golden section*.

6.2.10. The underestimation of the mathematical discovery of George Bergman

In mathematics, there is one "strange" tradition. Mathematicians tend to underestimate the mathematical achievements of some of their contemporaries (as it happened with *Lobachevsky's geometry*, which was strongly criticized by the official academic science of Russia) and overestimate the achievements of other mathematicians (as it happened with Cantor's set theory). Unfortunately, traditionally truly epochal mathematical discoveries are first subjected to harsh criticism and even ridicules from famous mathematicians, and only about 50 years later, as a rule, after the death of the authors of mathematical discoveries, new mathematical theories are recognized by the mathematical community and occupy a worthy place in mathematics. The examples of underestimation of the mathematical theories of Lobachevsky, Abel and Galois in the 19th century and their rehabilitation several decades later, when Lobachevsky, Abel and Galois were recognized as mathematical geniuses, are widely well known. The opposite example is the reappraisal of Cantor's theory of infinite sets at the initial stage and the recognition of this theory as a kind of "mathematical pathology" (Henri Poincaré) which caused the crisis in the foundations of modern mathematics.

Unfortunately, the 20th century did not become an exception from this tradition. A similar situation happened with the mathematical discovery of the young American mathematician, George Bergman. In 1957, George Bergman published the article "A numeral system with an irrational base" [54] in the American journal *Mathematics Magazine*. In this article, Bergman proposed a very unusual extension of the concept of a positional numeral system; this conception changes our ideas about numeral systems.

Unfortunately, Bergman's article [54] wasn't noticed at that time by either mathematicians, engineers, or experts in the field of computer technology. Journalists were surprised only by the fact that George Bergman made his outstanding mathematical discovery at the age of 12 years, and therefore, the article about the publication of the young mathematical talent of America was published in the American journal, *Mathematics Magazine*. But the mathematicians of that time, like Bergman himself, failed to appreciate the significance of this discovery for the development of mathematics and computer science. And only 23 years later, the significance of Bergman's discovery was appreciated in Stakhov's papers [58, 59] and 1984 book [19].

The strategic mistake of the 20th century mathematicians and experts in computer science was in the fact that they ignored Bergman's mathematical discovery, which can rightly be classified as the *greatest mathematical discovery in the field of numeral systems* (after the Babylonian positional numeral system with base 60, decimal and binary numeral systems). Bergman's system can lead to new computers and new number theory based on the *golden ratio* [85, 92, 97].

6.3. Beauty and Aesthetics of Harmony Mathematics

6.3.1. Hutcheson aesthetic principles

What is the beauty of science? In the rich history of world culture, there were researchers who dared to take up the analysis of the beauty of science. Perhaps, the first of them was the Scottish philosopher Francis Hutcheson (1694–1747), the author of the work *Studies on the Origin of Our Ideas of Beauty and Virtue in Two Treatises.* In the section "Beauty of Theorems", Hutcheson puts forward three basic principles of the beauty of science:

(1) the beauty is unity in diversity;
(2) the beauty lies in the universality of scientific truth;
(3) the scientific beauty is the acquisition of unobvious truth.

Alexander Voloshinov explains the essence of the *principle of unity in diversity* as follows [154]:

> *"...Any mathematical theorem contains in itself an infinite number of truths, which are valid for each specific object, but at the same time, all these specific truths are collected in one common truth, namely, in the theorem. For example, the Pythagorean Theorem is valid for an infinite set of concrete right-angled triangles, but all this variety of triangles has one common property, described by this theorem."*

We can say the same thing about the *task of the division of a segment in extreme and mean ratio* [32] called the *golden section.* There is an infinite number of segments, which can be divided into two unequal parts in the *golden section.* Moreover, all these divisions have a common property: the ratio of a larger segment to a smaller one is equal to the ratio of the whole segment to a larger one. This is an *example of unity in diversity.*

Let's now consider the second *Hutcheson principle,* the *principle of universality of scientific truth.* Hutcheson writes thus [154]:

> *"Theorems have another beauty that cannot be bypassed and which consists in the fact that one theorem can contain a huge quantities of consequences that are easily derived from it ... When we explore the Nature, a knowledge of certain great principles or general forces of Nature has a similar beauty, which brings to countless consequences. Such is the principle of the gravity in the scheme of Sir Isaac Newton...*
>
> *And we enjoy this pleasure, even if we don't have any prospects for obtaining any other benefit from this method of deduction, except for the direct pleasure from the contemplation of beauty."*

In any field of science, we can find a confirmation of this *Hutcheson principle*: in mathematics, this is the *Vieta theorem*, connecting the roots of algebraic equations with their coefficients; in physics, these are *Newton's laws* and *Maxwell's electromagnetism equations*; in biology, the *laws of genetics*.

Finally, Hutcheson's third aesthetic principle is the attainment of unobvious truth. The "2 × 2 = 4" theorem is an obvious truth and does not give us aesthetic pleasure. Voloshinov emphasizes as follows [154]:

> *"Only the discovery of truths, hidden from us by science or nature, the discovery, that requires efforts, that is, the knowledge of the unknown truth, gives us true pleasure at the end of the road."*

It is appropriate to recall here the famous statement of Heraclitus: *"Latent harmony is stronger than obvious."*

It is clear that all three aesthetic principles, derived by Hutcheson, are valid for any science, but, first of all, they are valid to mathematics. Mathematics has always been considered as the *queen of science*, and therefore, the *aesthetic principles of science are most clearly manifested in mathematics.*

6.3.2. Dirac's principle of mathematical beauty

Nowadays, no one doubts that *"science is beautiful, because it reflects and refracts in our consciousness the beauty, harmony and unity of the Universe"* [154]. But physicists went even further. The English physicist, Nobel Laureate Paul Dirac (1902–1984) put forward the thesis: "*The beauty is the criterion of the truth of physical theory.*" Dirac not only understood the beauty of the mathematical formulations of the theory but also understood the heuristic, regulatory role of the beauty as a methodological principle for the construction of scientific knowledge. Dirac remarks as follows [154]:

> "*I feel that a theory, if it is correct, should be beautiful, because we are guided by the principle of beauty, when we establishing fundamental laws. So, in studies, based on mathematics, we are often guided by the requirement of the mathematical beauty. If the equations of physics are ugly from a mathematical point of view, this means that they are imperfect and that the theory is flawed and needs to be improved. There are the cases when we should give a preference to mathematical beauty before agreement of theory with experiment. It is as if God created the Universe on the basis of excellent mathematics and we considered it reasonable to assume that the basic ideas should be expressed in terms of excellent mathematics.*"

6.4. Mathematics of Harmony from an Aesthetic Point of View

The outstanding English mathematician, philosopher and social activist Laureate of the Nobel Prize Bertrand Russell (1872–1970) expressed his relation to the beauty of mathematics in the following remarkable words:

> "*Mathematics possesses not only truth, but also the high beauty, the beauty refined and rigorous, sublimely pure, and striving for true perfection, which is characteristic only for the greatest examples of arts.*"

This statement can be attributed completely to the *Mathematics of Harmony* [6]; this three-volume book is devoted to the presentation of philosophical and historical principles of the *Mathematics of Harmony* [6].

It is interesting to track down, how the above-considered *Hutcheson's aesthetic principles* and *Dirac's principle of mathematical beauty* [154] are applicable to the *Mathematics of Harmony* [6].

Almost all the main mathematical results and geometric objects, obtained in the framework of the *Mathematics of Harmony* [6], including the *Fibonacci number theory* [8, 9, 11], are aesthetically perfect and evoke a sense of beauty and aesthetic pleasure. Let's recall only some of them.

6.4.1. Aesthetics of the golden section

In his famous *Elements* (13 books), Euclid already had introduced in Book II the *task of division of a segment into extreme and mean ratio*, this task is called the *golden section* in modern science. For many millennia, the *golden section* attracted admiration due to its amazing mathematical and geometric properties (Tables 6.1 and 6.2).

Table 6.2 provides the most famous geometric objects based on the *golden section*.

6.4.2. Aesthetics of Fibonacci and Lucas numbers

Fibonacci numbers had been introduced in the 13th century BC by the famous Italian mathematician Leonardo from Pisa (known by the nickname Fibonacci). Fibonacci advanced almost two centuries beyond the Western European mathematicians of his time. Similar to Pythagoras, who got his "scientific education" from the Egyptian and Babylonian priests and then made great contribution to the transmission of this knowledge into ancient Greek science, Fibonacci obtained his mathematical education in the Arabian educational

Table 6.1. Aesthetics of the "golden section".

1	Equation of the "golden section"	$x^2 - x - 1 = 0$
2	The golden proportion	$\Phi = \frac{1+\sqrt{5}}{2}$
3	The basic identities	$\Phi^n = \Phi^{n-1} + \Phi^{n-2} = \Phi \times \Phi^{n-1}$; $n = 0, \pm 1, \pm 2, \pm 3, \ldots$
4	Representation in the form of chain fraction	$\Phi = 1 + \cfrac{1}{1 + \cfrac{1}{1 + \cfrac{1}{1 + \cfrac{1}{1 + \cdots}}}}$
5	Representation in radicals	$\Phi = \sqrt{1 + \sqrt{1 + \sqrt{1 + \sqrt{1 + \cdots}}}}$

Table 6.2. Geometric objects based on the golden section.

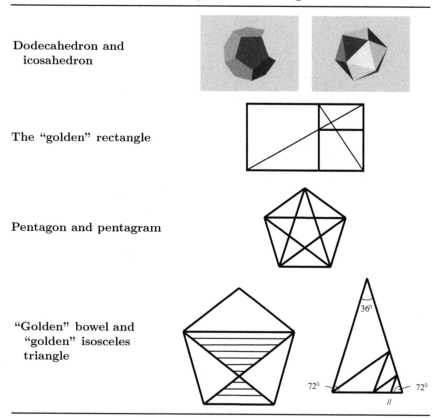

Dodecahedron and icosahedron	
The "golden" rectangle	
Pentagon and pentagram	
"Golden" bowel and "golden" isosceles triangle	

institutions and then he tried implementing many new scientific Arabian ideas into the Western European science and education. Similar to Pythagoras, the historical role of Fibonacci for the Western world was in the fact that through his mathematical books, Fibonacci tried to transmit the Arabian mathematical knowledge into the Western European science and in this way, he created the new foundations for further development of the European science, mathematics and education.

Table 6.3 shows a number of well-known mathematical properties of the Fibonacci and Lucas numbers obtained by well-known mathematicians in subsequent centuries. The aesthetic character of these results doesn't require much evidence.

Table 6.3.   Aesthetics of Fibonacci and Lucas numbers.

1 Recurrent relations for Fibonacci and Lucas numbers	$F_n = F_{n-1} + F_{n-2}$; $F_1 = F_2 = 1$ $L_n = L_{n-1} + L_{n-2}$; $L_1 = 1,\ L_2 = 3$ <table><tr><td>n</td><td>0</td><td>1</td><td>2</td><td>3</td><td>4</td><td>5</td><td>6</td></tr><tr><td>F_n</td><td>0</td><td>1</td><td>1</td><td>2</td><td>3</td><td>5</td><td>8</td></tr><tr><td>F_{-n}</td><td>0</td><td>1</td><td>-1</td><td>2</td><td>-3</td><td>5</td><td>-8</td></tr><tr><td>L_n</td><td>2</td><td>1</td><td>3</td><td>4</td><td>7</td><td>11</td><td>18</td></tr><tr><td>L_{-n}</td><td>2</td><td>-1</td><td>3</td><td>-4</td><td>7</td><td>-11</td><td>18</td></tr></table>
2 Extended Fibonacci and Lucas numbers	
3 Cassini's formula	$F_n^2 - F_{n-1}F_{n+1} = (-1)^{n+1}$, $n = 0, \pm1, \pm2, \pm3, \ldots$
4 Kepler's formula	$\displaystyle\lim_{n\to\infty} \frac{F_n}{F_{n-1}} = \Phi = \frac{1+\sqrt5}{2}$
5 Binet's formulas for Fibonacci and Lucas numbers	$F_n = \begin{cases} \dfrac{\Phi^n + \Phi^{-n}}{\sqrt5} & \text{for } n = 2k+1 \\[2mm] \dfrac{\Phi^n - \Phi^{-n}}{\sqrt5} & \text{for } n = 2k \end{cases}$ $L_n = \begin{cases} \Phi^n + \Phi^{-n} & \text{for } n = 2k \\[2mm] \Phi^n - \Phi^{-n} & \text{for } n = 2k+1 \end{cases}$
6 Hyperbolic Fibonacci and Lucas sines and cosines	$sFs = \dfrac{\Phi^x - \Phi^{-x}}{\sqrt5}$; $cFs = \dfrac{\Phi^x + \Phi^{-x}}{\sqrt5}$ $sLs = \Phi^x - \Phi^{-x}$; $cLs = \Phi^x + \Phi^{-x}$
7 Fibonacci matrices	$Q = \begin{pmatrix} 1 & 1 \\ 1 & 0 \end{pmatrix}$; $Q^n = \begin{pmatrix} F_{n+1} & F_n \\ F_n & F_{n-1} \end{pmatrix}$; $\det(Q^n) = (-1)^n$
8 "Golden" matrices	$Q_0(x) = \begin{pmatrix} cFs(2x+1) & sFs(2x) \\ sFs(2x) & cFs(2x-1) \end{pmatrix}$; $\det Q_0(x) = +1$ $Q_1(x) = \begin{pmatrix} sFs(2x+2) & cFs(2x+1) \\ cFs(2x+1) & sFs(2x) \end{pmatrix}$; $\det Q_0(x) = -1$

6.4.3. Aesthetics of Fibonacci and Lucas p-numbers ($p = 0, 1, 2, 3, \ldots$)

But really, the latest results in the field of *Mathematics of Harmony* [6], such as the *Fibonacci and Lucas p-numbers*, which are a generalization of the classical binary numbers ($p = 0$) and the *classical Fibonacci and Lucas numbers* ($p = 1$) and also the *golden p-proportions*, which are a generalization of the *classical golden proportion* ($p = 1$), produce great aesthetic impression (see Table 6.4).

Table 6.4. Aesthetics of Fibonacci p-numbers and "golden" p-proportions ($p = 0, 1, 2, 3, \ldots$).

	$\begin{array}{c} 1 \\ 1 \quad 1 \\ 1 \quad 2 \quad 1 \\ 1 \quad 3 \quad 3 \quad 1 \\ 1 \quad 4 \quad 6 \quad 4 \quad 1 \\ 1 \quad 5 \quad 10 \quad 10 \quad 5 \quad 1 \end{array}$

Fibonacci numbers as the diagonal sums of Pascal's triangle

$$\begin{array}{c} 1 \\ 1 \quad 1 \\ 1 \quad 2 \quad 1 \\ 1 \quad 3 \quad 3 \quad 1 \\ 1 \quad 4 \quad 6 \quad 4 \quad 1 \\ 1 \quad 5 \quad 10 \quad 10 \quad 5 \quad 1 \\ 1 \quad 6 \quad 15 \quad 20 \quad 15 \quad 6 \quad 1 \\ 1 \quad 7 \quad 21 \quad 35 \quad 35 \quad 21 \quad 7 \quad 1 \\ 1 \quad 8 \quad 28 \quad 56 \quad 70 \quad 56 \quad 28 \quad 8 \quad 1 \end{array}$$

$$\begin{array}{c} 1 \\ 1 \\ 2 \\ 3 \\ 5 \\ 8 \\ 13 \\ 21 \\ 34 \end{array}$$

Recurrent relation for Fibonacci p-numbers	$F_p(n) = F_p(n-1) + F_p(n-p-1)$ $F_p(1) = F_p(2) = \cdots = F_p(p+1) = 1$
Representation of Fibonacci p-numbers through binomial coefficients	$F_p(n+1) = C_n^0 + C_{n-p}^1 + C_{n-2p}^2 + C_{n-3p}^3 + \cdots$
Characteristic equation for the golden p-proportions	$x^{p+1} - x^p - 1 = 0 \ (p = 0, 1, 2, 3, \ldots) \rightarrow$ the positive root $x_1 = \Phi_p$ is the golden p-proportion
Cassini's formula for Fibonacci p-numbers	$\lim_{n \to \infty} \dfrac{F_p(n)}{F_p(n-1)} = \Phi_p$
Basic identities for the golden p-proportions	$\Phi_p^n = \Phi_p^{n-1} + \Phi_p^{n-p-1} = \Phi_p \times \Phi_p^{n-1}, \ n = 0, \pm 1, \pm 2, \pm 3, \ldots$ $p = 0: \ 2^n = 2^{n-1} + 2^{n-1} = 2 \times 2^{n-1}, \ n = 0, \pm 1, \pm 2, \pm 3, \ldots$ $p = 1: \ \Phi^n = \Phi^{n-1} + \Phi^{n-2} = \Phi \times \Phi^{n-1}, \ n = 0, \pm 1, \pm 2, \pm 3, \ldots$

6.4.4. Properties of Fibonacci and Lucas λ-numbers and "metallic proportions"

Mathematical identities for the *Fibonacci and Lucas λ-numbers*, which are a generalization of the classical *Fibonacci and Lucas numbers*, and the *metallic means* or *metallic proportions*, which are a generalization of the *classical "golden" proportion*, also produce great aesthetic impression (see Table 6.5).

Table 6.5. Aesthetics of Fibonacci and Lucas λ-numbers and "metallic proportions".

Recurrent relations for the Fibonacci and Lucas λ-numbers	$F_\lambda(n+2) = \lambda F_\lambda(n+1) + F_\lambda(n);\quad F_\lambda(0) = 0, F_\lambda(1) = 1$ $L_\lambda(n+2) = \lambda L_\lambda(n+1) + L_\lambda(n);\quad L_\lambda(0) = 2, L_\lambda(1) = \lambda$
Cassini's formulas for Fibonacci λ-numbers	$F_\lambda^2(n) - F_\lambda(n-1)F_\lambda(n+1) = (-1)^{n+1};$ $n = 0, \pm 1, \pm 2, \pm 3, \ldots$
Metallic proportions Φ_λ	$x^2 - \lambda x - 1 = 0 \to \Phi_\lambda = \frac{\lambda + \sqrt{4 + \lambda^2}}{2}$
Basic identity for the "metallic proportions"	$\Phi_\lambda^n = \lambda \Phi_\lambda^{n-1} + \Phi_\lambda^{n-2} = \Phi_\lambda \times \Phi_\lambda^{n-1}$
Representation in the form of chain fraction	$\Phi_\lambda = \lambda + \cfrac{1}{\lambda + \cfrac{1}{\lambda + \cfrac{1}{\lambda + \cdots}}}$
Representation in radicals	$\Phi_\lambda = \sqrt{1 + \lambda\sqrt{1 + \lambda\sqrt{1 + \lambda\sqrt{1 + \cdots}}}}$
Gazale's formulas	$F_\lambda(n) = \frac{\Phi_\lambda^n - (-1)^n \Phi_\lambda^{-n}}{\sqrt{4 + \lambda^2}};\quad L_\lambda(n) = \Phi_\lambda^n + (-1)^n \Phi_\lambda^{-n}$
Hyperbolic Fibonacci and Lucas λ-sines and cosines	$sF_\lambda(x) = \frac{\Phi_\lambda^x - \Phi_\lambda^{-x}}{\sqrt{4 + \lambda^2}};\quad cF_\lambda(x) = \frac{\Phi_\lambda^x + \Phi_\lambda^{-x}}{\sqrt{4 + \lambda^2}}$ $sL_\lambda(x) = \Phi_\lambda^x - \Phi_\lambda^{-x};\quad cL_\lambda(x) = \Phi_\lambda^x + \Phi_\lambda^{-x}$

Thus, as it follows from the analysis of mathematical formulas and geometric objects, given in Tables 6.1–6.5, the *Mathematics of Harmony* [6], which was created by many generations of outstanding scientists and thinkers, starting from the ancient period, undoubtedly has high perfection and exceptional aesthetic properties.

Dirac's principle of mathematical beauty [154] in connection with the *Mathematics of Harmony* [6] is already embodied in the outstanding modern scientific discoveries, based on the mathematical results, obtained in ancient science. Two of them (*quasi-crystals* and *fullerenes*) have already been awarded with the Nobel Prize.

Chapter 7

Epilogue

7.1. A Brief History of the Concept of Universe Harmony

Differentiation of modern science and its division into separate areas often do not allow one to view the general picture of science and the main trends of its development. However, in science, starting from the ancient times, there are research objects that unite the disparate scientific facts into a single whole. *Platonic solids* and the *golden section* can be attributed to such fundamental geometric objects. The ancient Greeks elevated these geometric objects to the level of *the main objects, which express the Universal Harmony.*

This three-volume book is devoted to the discussion of one the oldest scientific conceptions of ancient mathematics: the conception of Universe Harmony. The greatest theoretical contribution to the development of this conception in the ancient period was made by Pythagoras, Plato, and Euclid, who can be considered as the greatest geniuses of ancient science.

The most interesting point consists in the fact that the interest in such scientific conceptions of ancient science not only doesn't decrease but also increases even more in the process of further development of science. For centuries or even millennia, starting with Pythagoras, Plato, and Euclid, these conceptions were the subject of admiration and worship for some eminent personalities: During the

Renaissance — Leonardo da Vinci, Luca Pacholi, Johannes Kepler; in the 19th century — Zeising, Lucas, Binet, and Klein. In the 20th century, the interest in these geometric objects (the *golden section* and *Platonic solids*), which expressed *Universal Harmony* in the ancient period, increased significantly in modern mathematics, thanks to the studies of the Soviet mathematician, Nikolay Vorobyov and the American mathematician, Verner Hoggatt, from whose works the process of *harmonization of mathematics* begins [68]. The development of this direction led to the creation of *Mathematics of Harmony* as a new interdisciplinary direction and the *"golden"* *paradigm* of modern science [6].

The latest theoretical discoveries in the various fields of modern science, based on the *Platonic solids*, the *golden section, Fibonacci numbers*, new scientific results, obtained in the framework of *Mathematics of Harmony* [6], and new discoveries in this area, namely, *new geometric theory of phyllotaxis, general theory of hyperbolic functions, algorithmic measurement theory, Fibonacci and golden ratio codes*, the *"golden" number theory, Fibonacci microprocessors, Fibonacci computers*, and so on, create a general picture of the movement of modern science toward the *"golden" scientific revolution*, which is one of the characteristic trends in the development of modern science.

This conclusion is confirmed by a fairly impressive list of widely known scientific events in the contemporary history of science (the creation of the *American Fibonacci Association* and the *Slavonic Golden Group*) as well as articles and books [1–100], published in well-known scientific journals and by the world famous publishing houses (including World Scientific), and also by the outstanding scientific discoveries of modern science, whose authors were awarded Nobel Prizes (*quasi-crystals, fullerenes*). These scientific discoveries provide the confirmation of the fact that the ancient problems of *Harmony, "golden section", and Platonic solids* naturally find place in modern science, mathematics and education, and become a fruitful source of new scientific discoveries in mathematics, computer science and natural sciences.

7.2. More on the Doctrine of Pythagoreanism, Pythagorean MATHEMs, and Pythagorean Mathematical and Scientific Knowledge

The concept of Harmony is one of the central notions of *Pythagoreanism*, the main Pythagorean scientific doctrine (see Wikipedia paper *Pythagoreanism*). As mentioned in the Preface, according to the tradition, the *Pythagoreans* were divided into two separate schools of thought: the *mathematikoi* (*mathematicians*) and the *akousmatikoi* (*listeners*). The *listeners* had developed the religious and ritual aspects of the *Pythagoreanism;* the *mathematicians* studied four Pythagorean MATHEMs: *arithmetics, geometry, spherics* and *harmonics*. These MATHEMs, according to Pythagoras, were the main composite parts of mathematics. Unfortunately, **the Pythagorean MATHEM of** *harmonics* **was lost in mathematics in the process of its historical development.** At least, basic conceptions of *Harmonics* such as *golden section* and *Platonic solids* do not play in modern mathematics such a *"key"* or *fundamental* role, which they played in the mathematics of Pythagoras, Plato and Euclid.

Pythagoras is mostly remembered for his mathematical ideas such as the definition of *natural numbers, Pythagorean Theorem,* and creation of mathematical methods, such as *arithmetic* and *geometry*. The *mathēmatikoi* (*mathematicians*) claimed that the *natural numbers* were at the heart of everything and *Pythagorean mathematicians* constructed a *new view on the Cosmos*. According to the *Pythagorean tradition*, Earth was removed from the center of the Universe. The *mathēmatikoi* (*mathematicians*) believed that Earth, along with other celestial solids, orbited around a central fire. They also believed that this resembles a celestial *Harmony*.

7.2.1. The most famous numerical achievements of Pythagoreans

Creating the foundations of the *elementary theory of numbers* was one of the most important achievements of *Pythagorean arithmetic*.

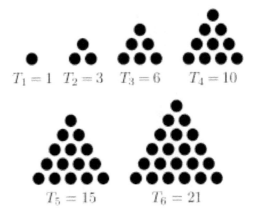

Fig. 7.1. The first six triangular numbers.

They considered the *natural numbers* as the main subject of the *elementary theory of numbers*. The *Pythagorean mathematicians* represented numbers graphically, not symbolically through letters. Pythagoreans used dots, also known as *psiphi* (pebbles), to represent *natural numbers* in the form of *triangles*, *squares*, *rectangles*, and *pentagons*. Figure 7.1 demonstrates the representation of the so-called *triangular numbers* 1, 3, 6, 10, 15, 21 and so on by using dots presented in *triangular form*.

This promoted a visual comprehension of mathematics and allowed a geometrical exploration of numerical relationships. Pythagorean mathematician studied the so-called *perfect numbers*, which were equal to the sum of all their divisors. For example, $28 = 1 + 2 + 4 + 7 + 14$. The theory of the *odd* $(2, 4, 6, \ldots)$ and *even* $(3, 5, 7, \ldots)$ *natural numbers* was central to the *Pythagorean arithmetic*.

7.2.2. Geometric achievements of Pythagoreans

Pythagorean philosophers believed that there was a close relationship between *numbers* and their *geometrical forms* (see Fig. 7.1). Early Pythagorean philosophers proved simple geometrical theorems, including *the sum of the angles of a triangle equals two right angles*. Pythagoreans also came up to three of the five regular polyhedra:

the *tetrahedron*, the *cube*, and the *dodecahedron*. The sides of a *regular dodecahedron* are *regular pentagons*, which for Pythagoreans symbolized *health*. They also revered the *pentagram* because each diagonal divides the two others in the *golden ratio*.

7.2.3. Pythagorean theory of music

Pythagoras was a pioneer in the mathematical and experimental study of music [10]. He discovered quantitative mathematical relationships of music through arithmetic ratios. Pythagoras is credited with the discovery of the fact that the most harmonious musical intervals are created by the simple numerical ratios of the first four integer numbers which derive respectively from the relations of string length: the *eighth* $(1/2)$, the *fifth* $(2/3)$ and the *fourth* $(3/4)$. The sum of those numbers $1 + 2 + 3 + 4 = 10$ (*tetraxis*) was for Pythagoreans the *perfect number* because it contained in itself *the whole essential nature of numbers*. Werner Heisenberg referred to this musical arithmetic as *among the most powerful advances of human science* because it enabled one to measure sounds in space.

7.2.4. Pythagorean numerical Harmony

For Pythagorean philosophers, the basic property of numbers was expressed in the harmonious interplay of opposite pairs. The *Harmony* assured the balance of the opposite forces. Pythagoras in his doctrines named *natural numbers* and *their symmetries* as the *first principle* and called these numerical symmetries *Harmony*. Pythagorean philosophers believed that numbers were the elements of the whole existing world, but the Universe on the whole consists of *Harmony* and *Numbers*.

Pythagoreans believed in *musica universalis* [10]. They believed that the stars must produce sounds in space because they were large swiftly moving solids. The early Pythagorean philosopher Philolaus (470–385 BC) argued that the structure of cosmos was determined by the *musical numerical proportions* of diatonic octave, which contained the fifth and fourth harmonic intervals.

7.3. Mathematization of Harmony and Harmonization of Mathematics

7.3.1. A brief history

These are two important mathematical concepts introduced by the author in Ref. [68]. They are directly related to the history of science and mathematics and deserve to be mentioned in the final part of this book. Of the latest publications on the website of the Academy of Trinitarism, the author was most impressed by the article "About the past and future gods, priests and prophets of science" [152], written by Denis Kleschev, the young Russian historian of mathematics and science.

Earlier, the author was almost equally impressed by two publications: the book by the Belarusian philosopher Edward Soroko, *Structural Harmony of Systems* [4], published in 1984, and the book by the American mathematician Morris Klein *Mathematics. The Loss of Certainty* [101] translated into Russian in 1984. These books have a revolutionary significance for the development of modern science and mathematics.

The 1984 book by Edward Soroko [4] attracted the attention of the scientific community to one of the most important trends in the development of modern science: the revival of the *Pythagorean doctrine about the numerical harmony of the Universe*. The 1984 book by Morris Klein [101] sank mathematicians in deep thought over the deplorable state of modern mathematics, after the emergence of a new crisis in the foundations of mathematics, connected with the paradoxes in Cantor's theory of sets. About this, Morris Klein wrote the following [101]:

> *"The disagreements over the foundations of the most "unshakable" of the sciences (mathematics) caused surprise and disappointment...*
> *The current state of mathematics is nothing more than a pathetic parody on mathematics of the past with its deep-rooted and widely known reputation of the impeccable ideal of truth and logical perfection."*

It is curious to note that the year 1984 is a special, crucial, very productive year for the *golden section* and its applications.

Let's consider only some examples. On November 12, 1984, in a small article, published in the authoritative journal *Physical Review Letters*, the experimental evidence was given for the existence of the metallic alloy with exceptional properties (the author of the discovery is the Israeli physicist Dan Shechtman). The crystallized structure of this alloy had the "icosahedral" symmetry, that is, the fifth-order symmetry, which is strictly prohibited by classical crystallography. Alloys with such unusual properties were named *quasi-crystals*. **Thanks to this discovery, the *golden section*, which lies at the basis of the icosahedron, dodecahedron and the fifth-order symmetry (pentagram), came to the forefront of modern physics.**

In Gratia's article "Quasi-crystals" (*Advances in Physical Sciences*, 1988) [115], devoted to this discovery, it is noted that *its value in the mineral world can be put in one row with the addition of the concept of irrational numbers in mathematics*. The year 2011 became a landmark year not only for the quasi-crystals but also for all *harmonious mathematics*. For this discovery, Dan Shechtman was awarded the Nobel Prize in Chemistry, that is, the problem of the *golden section*, described in Euclid's *Elements* (Proposition II.11), suddenly became the subject of discussion in the Nobel Committee!

The discovery of the quasi-crystals (Shechtman) was a worthy gift to the 100th anniversary of the publication of the book of the German mathematician Felix Klein *The Icosahedron and The Solution of Equations of the 5th Degree* (1884) [113], in which 100 years before the publication of Shechtman's article, the role of the icosahedron in the development of modern science, in particular, of contemporary mathematics was predicted. **The Nobel Prize for the discovery of the quasi-crystals is a brilliant confirmation of the fact that the *golden section* is indeed a fundamental constant of Nature.**

Note also that in 1984, the American Fibonacci Association held its *First International Conference on Fibonacci numbers and Their Applications*. Starting from this year, this famous mathematical conference had been held regularly (once every 2 years).

It can be stated with sufficient confidence that the books by Edward Soroko [4] and Alexey Stakhov [19], published in 1984, became the beginning of a new stage in the development of the "golden section" movement in the world because since this year, the "golden section" movement began to be developed actively by the members of the so-called *Slavic Golden Group*, an informal association of Slavic scientists (Ukraine, Russia, Belarus, Poland). The *Slavic Golden Group* was organized in Kiev in 1992 during the First International Seminar *The Golden Section and Problems of System's Harmony*.

But let us return again to the article by Denis Kleshchev [152]. In this article, he formulated the following two important statements:

"(1) *A paradigm shift in mathematics precedes the paradigm shift in physics; this historical regularity refutes the stereotype, that has emerged in our days, that mathematics and physics are not be interconnected methodologically. Such a connection exists, and it is not accidental that after the monograph on the history of mathematics "Mathematics. Loss of Certainty" [101] Morris Klein wrote another book, "Mathematics and the search for Knowledge" [153], dedicated to a similar process of the loss of certainty in physics.*

(2) *The concept of Alexey Stakhov, who upholds the indisputable for all historians of mathematics (and extremely unpleasant for modern priests of science) thing, is closest to such views on the future of mathematics, namely the fact that mathematics originated in the inseparable interrelation of three main problems:* **measurement problem** *(geometry, trigonometry),* **counting problem** *(arithmetic, numeral systems), and also* **harmony problem** *(systematization and generalization of knowledge). Although the evidence of Proclus Diadoch (V century AD), which reveals the meaning of the Euclidean Elements, compiled as a strictly axiomatic construction of the five regular polyhedrons, has a lot of evidence in ancient science, modern mathematicians continue to hide this historical fact from us.*"

In the article [68], published in 2011, the author compared the two opposite processes of mathematics development. The first process was characteristic of ancient mathematics (in particular, for the ancient Greek mathematics of Pythagoras, Plato and Euclid) and the second process is characteristic of contemporary mathematics.

Based on the *Proclus hypothesis* [95], which led to a new look at the Euclidean *Elements* and the history of mathematics after Euclid, the author came to the conclusion that in ancient Greece the science and mathematics developed under the flag of *Mathematization of Harmony*, and this idea was embodied most vividly in the Euclidean *Elements*.

At the same time, by analyzing the process of the development of contemporary science and mathematics and by using some sensational discoveries in modern mathematics, computer science and theoretical natural sciences (*Bergman's numeral system* [54], *Stakhov's codes of the golden proportion* [19], *Soroko's law of structural harmony of systems* [4], *Bodnar's geometry of phyllotaxis* [42], *fullerenes* (Nobel Prize in Chemistry, 1996) [116], *Shechtman's quasi-crystals* (Nobel Prize in Chemistry, 2011) [115], and so on), the author came to the conclusion that the process of *Harmonization of Mathematics* prevails in the development of modern science and mathematics.

This conclusion is confirmed by the imposing number of modern publications on this topic, in particular, by the publication of Stakhov's 2009 book *The Mathematics of Harmony. From Euclid to Contemporary Mathematics and Computer Science* in one of the most prestigious international publishing houses "World Scientific" [6], the book *Harmony. A new way of looking at our world* (2010), written by the Prince of Wales with Tony Juniper and Jan Skelly [49] and the book *Mathematics and the History of the Golden Section* by the Armenian philosopher and physicist, Hrant Arakelian (2014) [50].

From the above consideration, the following conclusion about the deep relationship between the two most important processes that took place in science and mathematics more than 2000 years ago and that takes place currently: the process of *Mathematization of Harmony*, which began in ancient Greece in the sixth–fifth centuries BC (mathematics of Pythagoras and Plato) and ended in the third century BC with writing of the most famous mathematical work of the ancient era, the Euclidean *Elements,* and the process of the *Harmonization of Mathematics,* which began in the second half of the 20th century (the works of Nikolay Vorobyev [8], Verner Hoggatt [9], Stefan Vaida [11] and other Fibonacci mathematicians) and continues to the present days (the books by Alexey Stakhov [6], the Prince of Wales with Tony Juniper and Jan Skelly [49] and Hrant Arakelian [50]).

7.4. *The Structure of Scientific Revolutions* by Thomas Kuhn

Thomas Samuel Kuhn (1922–1996) was the prominent American philosopher of science, whose 1962 book *The Structure of Scientific Revolutions* [139] influenced both academic and popular circles; he introduced the term *paradigm shift*, which is the English-language notion of the *Golden Revolution*.

The basic ideas of the Kuhn 1962 book *The Structure of Scientific Revolutions* [139] consists in the following. Kuhn suggested several interesting hypotheses concerning the progress of scientific knowledge. **The most important of them consisted in the fact that scientific knowledge undergoes periodic *"paradigm shifts"*, which open up new approaches to understanding what scientists would never have considered true before, and therefore, the notion of scientific truth, at any given moment, cannot be established solely by objective criteria but is defined by a consensus of the scientific community.** The competing paradigms are frequently incompatible models of reality. Thus, our comprehension of science can never rely wholly upon "objectivity" alone. Science must take into consideration subjective points of view as well because all objective conclusions are ultimately founded upon the subjective world outlook of its researchers and participants.

Thomas Kuhn's book *The Structure of Scientific Revolutions* (1962) [139] is devoted to the analysis of the *history of science*. The main goal of Kuhn's book is to prove the fact that science is developed "stepwise" through *scientific revolutions*. *Scientific revolution* means change of scientific paradigm in science. Such development makes sense only within the framework of a particular paradigm, the historically established system of views.

Thomas Kuhn's book received wide resonance in modern scientific community. The book became a kind of sensation among researchers in various areas of science and among historians of science. The book *The Structure of Scientific Revolutions* is one of the most cited scientific books in the entire history of science.

This three-volume book *Mathematics of Harmony as a New Interdisciplinary Direction and "Golden" Paradigm of Modern Science* is a continuation of the author's 2009 book *The Mathematics of Harmony. From Euclid to Contemporary Mathematics and Computer Science* [6]. Its distinctive feature is to consider the scientific results, obtained in the book [6] from the point of view of Kuhn's book *The Structure of Scientific Revolutions* and to formulate *key findings* and *basic conclusions* from the present study, which allow one to treat a totality of new scientific results obtained in this three-volume book as the *"Golden" Paradigm of Modern Science*.

7.4.1. Kuhn's criteria to Theory Choice

In the essay, called *Objectivity, Value Judgment, and Theory Choice*, Kuhn formulates the *five principles*, taken from the penultimate chapter of his *Structure of Scientific Revolutions* [139], which help determine the validness of Theory Choice:

(1) *Accurateness principle*: The theory empirically must be adequate to *experimentations* and *observations*.
(2) *Consistency principle*: The theory internally and externally must be consistent with other theories.
(3) *Broad Scope principle*: The theory's consequences should extend beyond those that were initially used to explain.
(4) *Simplicity principle*: The theory has the simplest explanation, principally similar to *Occam's razor*.
(5) *Fruitfulness principle*: The theory must disclose new phenomena or new relationships between phenomena.

7.5. Main Conclusions and New Challenges

The main conclusions of this three-volume book from the Mathematics of Harmony as the "Golden" Paradigm of Modern Science reduce to the following:

(1) **The first conclusion** concerns the very concept of *Harmony*. This conclusion follows from researches stated in Vol. I. The widespread manifestation of the *golden section* and the *Fibonacci numbers* in natural objects (*the phenomena of phyllotaxis, "golden" spirals, pentagonal symmetry*, etc.), as well as modern scientific discoveries in various fields of natural sciences, in particular, in chemistry (*fullerenes, Fibonacci regularity in the Periodic Table of Mendeleev*), crystallography (*quasi-crystals*), botany (*Bodnar's geometry of phyllotaxis*), genetics (*symmetric properties of the genetic code and "golden" gene matrixes*), quantum physics (*the symmetry of the quantum world* based on the *golden section*) and other areas of natural sciences are reasons for the fact that **Harmony is an objective scientific concept, which exists independent of our consciousness and is expressed in the harmonious organization of all things, including Cosmos and Micro world.** But if *Harmony* is the objective concept, it must be the subject of mathematical research which is one of the main goals of this three-volume book.

(2) **The second conclusion** concerns the question of the fundamental role of *harmonious ideas* in the history of ancient mathematics. According to the *Proclus hypothesis* [95], the Euclidean *Elements*, the greatest mathematical work of the ancient era, was created under the direct influence of the *harmonious ideas* of Pythagoras and Plato. The main aim, which Euclid pursues in his *Elements,* was the creation of the complete geometric theory of *Platonic solids* (Book XIII), associated in ancient science with the *numerical Harmony of the Universe*. Thus, starting from the ancient Greek period, two mathematical directions, the *Classical Mathematics* and the *Mathematics of Harmony*, began developing in parallel and independent of each other. Both of these directions originated from one and the same source: the Euclidean *Elements*. At the

same time, *Classical Mathematics* borrowed from the *Elements* the following mathematical achievements: *geometric axioms (Euclidean axioms), principles of algebra, elementary theory of numbers, theory of irrationality,* and other mathematical achievements of ancient Greek mathematics, whereas *Mathematics of Harmony* borrowed from the Euclidean *Elements,* first of all, *the task of dividing a segment in extreme and mean ratio* (Proposition II.11), later called the *golden section,* and the geometric theory of the *regular polyhedra* (Book XIII) called *Platonic solids* [68]. At present, the process of the *harmonization of mathematics* [68] has been developing, which means returning to those ideas (*mathematization of harmony*), which according to the *Proclus hypothesis,* the ancient Greek mathematicians used for the creation of the first in the history of science *mathematical doctrine of Nature* [101]. The most important milestone in this process was the writing of the Euclidean *Elements.*

(3) **The third conclusion** concerns the place, which the *Mathematics of Harmony* occupies in the system of mathematical disciplines. The answer to this question is given by the prominent Ukrainian mathematician academician Yuri Mitropolskiy in his feedback [162]:

> *"One may wonder what place in the general theory of mathematics is occupied by Mathematics of Harmony created by Prof. Stakhov? It seems to me, that in the last centuries, as Nikolay Lobachevsky said, "mathematicians turned all their attention to the* ***Advanced Parts of Analytics, neglecting the origins of Mathematics and not willing to dig the field that already been harvested by them and left behind."*** *As a result, this created a gap between "Elementary Mathematics", the basis of modern mathematical education, and "Advanced Mathematics". In my opinion, the Mathematics of Harmony, developed by Prof. Stakhov, fills up that gap. I.e., the Mathematics of Harmony is a big theoretical contribution, first of all in the development of the "Elementary Mathematics" and as such should be considered of great importance for mathematical education."*

It follows from Mitropolsky's statement that the *Mathematics of Harmony* is directly related to such ancient mathematical theories as *measurement theory* and *elementary theory of numbers.*

(4) **The fourth conclusion** concerns the connection between the *Mathematics of Harmony* [6] and the *Theory of Measurement* [16, 17]. As it is known, the *Theory of Measurement* began developing in mathematics after the discovery of the *incommensurable segments*, based on the so-called *Eudoxus exhaustion method*, which allowed overcoming the *first crisis in the foundations of mathematics*, connected with the discovery of *incommensurability*. This method was the basis of the *Eudoxus–Archimedes axiom*, also called the *axiom of measurement*, one of the most important axioms in the system of *continuity axioms*. In the second half of the 19th century, *Cantor's axiom* was included to the *axioms of continuity*. With the help of the *Eudoxus–Archimedes axiom* and *Cantor's axiom*, it seemed possible to complete the creation of the *mathematical measurement theory*.

However, the joy was premature. After the discovery of paradoxes in the *Cantor theory of infinite sets*, great doubts arose about the validity of the usage of this axiom in mathematics [129] because *Cantor's axiom* is based on the abstraction of *actual infinity* which, according to Aristotle, is unacceptable in mathematics. The basis of the new mathematical theory of measurement, called the *Algorithmic Theory of Measurement* [16, 17], was the approach, adopted in *constructive mathematics*, which excludes the usage of the *abstraction of actual infinity* as an internally contradictory concept and takes the *abstraction of potential infinity*. For the first time, such an idea was expressed in the author's book *Introduction to the Algorithmic Theory of Measurement* [16]. Within the framework of the *algorithmic measurement theory* [16], the problem, which had never been solved in mathematics before, the problem of synthesizing *optimal measurement algorithms* was formulated. By using the *constructive idea of potential feasibility* and the *principle of measurement asymmetry* [57], the author obtained a number of unexpected mathematical results, in particular, he synthesized a new class of optimal measurement algorithms, based on the *Fibonacci p-numbers* and called *Fibonacci measurement algorithms*.

(5) **The fifth conclusion** concerns another ancient mathematical theory: the *elementary theory of numbers*. In this part, new

mathematical results were obtained: the *Fibonacci p-numbers* as a new class of recurrent numerical sequences, arising from the *Pascal triangle* (the "diagonal sums" of *Pascal's triangle*) as well as the *number system with an irrational basis* (the *golden proportion*), proposed in 1957 by the young American mathematician George Bergman [54], and the *codes of the golden proportion*, proposed by the author in 1984 [19]. These binary positional numeral systems change our ideas about numeral systems. A special class of irrational numbers, which includes the classic *golden proportion* and its generalization, *the golden p-proportions* are used as the basis of the new numeral systems. These numeral systems underlie the *"golden" theory of numbers* [53]. In the framework of the *"golden" theory of numbers*, new properties of natural numbers (*Z*- and *D*-properties, *F*-code, *L*-code, etc.) were discovered. **All the above mathematical results are part of the *"golden" elementary theory of numbers*.**

(6) **The sixth conclusion** concerns the connection of *Mathematics of Harmony* with the *hyperbolic geometry* and the *botanical phenomenon of phyllotaxis*. One of the unexpected mathematical results, obtained in the framework of *Mathematics of Harmony* is the *Fibonacci and Lucas hyperbolic functions* (the Ukrainian scientists Alexey Stakhov, Ivan Tkachenko, and Boris Rozin) based on the so-called *Binet formulas* [64, 65, 75]. These functions are a generalization of the *Binet formulas* to a continuous domain, and their theory is an extension of the theory of the *Fibonacci and Lucas numbers* to a continuous domain. With the introduction of the *hyperbolic Fibonacci and Lucas functions*, the *classical Fibonacci and Lucas number theory* seems to be "degenerated", because all the identities for the *Fibonacci and Lucas numbers* can be easily obtained from the corresponding identities for the *hyperbolic Fibonacci and Lucas functions* by using the elementary formulas, connecting these functions with the Fibonacci and Lucas numbers.

(7) **The seventh conclusion** concerns the generalized theories of the *golden section*, proposed and developed in the framework of the *Mathematics of Harmony* [6]. Here, the fundamental role is played by the ideas of the Armenian philosopher and physicist Grant Arakelian,

set forth in his outstanding article *On the World of Harmony, theory of the Golden Section and its Generalizations* [161]. The main Arakelian idea is that any theory that claims to be a generalization of the classical *golden section theory* or the classical *Fibonacci numbers theory* must satisfy certain *generic features* that unite classical theory with its generalization. From this point of view, the most interesting *generalized theories of the golden section* and the *Fibonacci numbers*, developed in the framework of the *Mathematics of Harmony* [6], include two mathematical theories. The first one is the theory of the *Fibonacci p-numbers* and the *golden p-proportions* (where $p = 0, 1, 2, 3, \ldots$ is a given nonnegative integer). The second one is the theory of the *Fibonacci λ-numbers*, developed in Gazale's book [34], and Spinadel's *metallic λ-proportions* (where $\lambda > 0$ is a given real number), introduced in Spinadel's article [33]. These generalized theories of the classical *Fibonacci numbers* and the classical *golden section* have pronounced *generic features*, characteristic of both classical *Fibonacci numbers* and the classical *golden proportion*. For the *Fibonacci p-numbers*, such *generic attribute* is the *Pascal triangle*, whose *diagonal sums* coincide with the *Fibonacci p-numbers*; the *generic attribute* for *golden p-proportions* is the *Kepler formula* (the limit of the ratio of the neighboring *Fibonacci numbers* tends to the classical *golden proportion*, while the limit of the ratio of the *neighboring Fibonacci p-numbers* tends to the *golden p-proportion*). For the *Fibonacci λ-numbers*, such a generic attribute is the *Cassini formula*, which is valid for both classical *Fibonacci numbers* and *Fibonacci λ-numbers*. Finally, the *generic attribute* for *Spinadel's metallic proportions* is the *rule of the mantissa conservation* established by Grant Arakelian [161].

(8) **The eighth conclusion** concerns the connection between the *Mathematics of Harmony* [6] and *Hilbert's Problems*, in particular, with *Hilbert's 10th and 4th Problems*. As it is known, *Hilbert's 10th problem* was solved in 1970 by the young Soviet mathematician Yuri Matiyasevich [150]. In Matiyasevich's work [150], it was stressed repeatedly that the solution to this problem was obtained by using the latest mathematical results in the field of *Fibonacci number*

theory, published by the eminent Soviet mathematician Nikolay Vorobyov in the third edition of his remarkable book *Fibonacci Numbers* (1969). *Hilbert's Fourth problem* was solved by Alexey Stakhov and Samuil Aranson in the first decade of the 21st century. An indirect confirmation of the recognition of this mathematical result by the world mathematical community is its publication in the first three issues of the international journal, *Applied Mathematics*. This solution is based on the use of the so-called *hyperbolic Fibonacci and Lucas λ-functions* introduced by Alexey Stakhov in 2006 [12].

(9) **The ninth conclusion** relates to the new scientific problem that *Mathematics of Harmony* [6] poses before the theoretical natural sciences. This problem follows directly from the *Bodnar geometry* [28], from the theory of the hyperbolic Fibonacci and Lucas functions [88–90] and finally from the complete solution of *Hilbert's Fourth Problem* [151]. The *Bodnar geometry* showed that the *world of phyllotaxis*, one of the most amazing phenomena of botany, is the *hyperbolic world*, based on the *hyperbolic Fibonacci and Lucas functions*, the basis of which is the classic *golden proportion*. Thus, in the botanical phenomenon of phyllotaxis, *hyperbolicity* manifests itself in the *gold*. This *Bodnar hypothesis* proved to be very fruitful and led to the creation of a *new geometric theory of phyllotaxis*, which has an interdisciplinary significance. However, the *Fibonacci and Lucas hyperbolic functions* are a particular case of *hyperbolic Fibonacci and Lucas λ-functions*. The latter functions are based on *Spinadel's metallic proportions*, in particular, on the *silver, bronze, copper*, and other types of *metallic proportions*. In this regard, the following fundamental problem, which can be set before theoretical natural sciences, follows from the above-mentioned mathematical results. It can be assumed that other types of *hyperbolic Fibonacci and Lucas λ-functions*, which are based on the *Spinadel's metallic proportions* and underlie at the solutions of Hilbert's Fourth problem, can become the basis for modeling new *hyperbolic worlds*, which may actually exist in Nature but which modern science has not yet discovered because it has not been known for the *hyperbolic Fibonacci and Lucas λ-functions* and no one put such a task

before modern science. Based on the brilliant success of the *Bodnar geometry* [28, 42], we can pose to theoretical physics, chemistry, crystallography, botany, biology and other branches of theoretical natural sciences the *challenge of finding the new hyperbolic worlds of nature based on the hyperbolic Fibonacci and Lucas λ-functions.*

It is quite possible that the role of the first of these functions, which can lead to the discovery of a *new hyperbolic worlds of Nature*, can be played by *the golden hyperbolic functions*, based on the *classical golden ratio*, and the *silver hyperbolic functions*, the basis of which is the new mathematical constant, the *silver proportion* $\Phi_2 = 1 + \sqrt{2}$, generated by the *relic number* $\sqrt{2}$, widely used in the models of contemporary theoretical natural sciences. According to the Russian researcher Alexander Tatarenko, the introduction of the *silver ratio* $\Phi_2 = 1 + \sqrt{2}$ into contemporary science is *the most important scientific breakthrough on the path to Truth about the Harmony of the World, comparable to the change of the Ptolemaic geocentrism to the Copernican helio-system* [141].

(10) **The 10th conclusion** concerns the beauty and aesthetics of *Mathematics of Harmony* [6]. All mathematical objects of *Mathematics of Harmony*, starting with the *classical golden ratio, Fibonacci numbers* and geometric objects based on them (*pentagram*, "*golden*" *rectangle*, "*golden*" *spiral*, "*golden*" *triangles*, etc.) and ending with the newest mathematical results in this field (*Fibonacci p-numbers, golden p-sections, Spinadel's metallic proportions, Fibonacci, and Lucas hyperbolic functions* and their generalizations, the *Fibonacci and Lucas λ-hyperbolic functions*) have beautiful and aesthetic perfection. The *aesthetic principles of Hutcheson* and *Dirac principles of mathematical beauty* [128, 129] are applicable to them; this is the ground to assert that the mathematical objects of *Mathematics of Harmony* [6] can be widely used in theoretical natural sciences to create new models of natural objects and structures.

(11) **The 11th conclusion** concerns the role of *Mathematics of Harmony* [6] in the development of computer science based on the classical *binary system.* As the outstanding Russian computer science

expert, Academician Yaroslav Khetagurov emphasizes, the use of modern microprocessors with low information reliability is a kind of *Trojan horse* (*zero redundancy*) that can lead to unpredictable consequences in modern applications of microprocessors. One of such areas is rocket and space technology, in which a number of catastrophes have already occurred due to the failures in the digital missile control system. Thus, from the foregoing facts, there follows a far from optimistic conclusion. **Mankind is becoming hostage to the classical binary numeral system, which is the basis of modern microprocessors and information technology.** Therefore, further development of information technology based on the *binary system* must be recognized as the *unacceptable direction for certain areas of applications.* The *binary system* cannot serve as the informational and arithmetic foundation for specialized computer and measurement systems (*space, transport control, and complex technological objects*) as well as *nano-electronic systems*, where the **problems of reliability, noise immunity, control, stability, survivability of systems come to the fore.** *Mathematics of Harmony* suggests a very elegant solution to these problems, which unexpectedly manifested itself in some critical applications of modern computers and has already become a *national problem* of some countries (Russia and the USA). The essence of this solution is to use the redundant numeral systems: *Bergman system* [54], *codes of the golden p-proportions* [19], *Fibonacci p-codes* [55] as new redundant positional numeral systems, which can become an alternative to the classical binary system for the mission-critical applications [56]. These new numeral systems are "binary", that is, they use the binary alphabet {0, 1} to represent numerical information. They retain all the well-known advantages of the classical binary system, but at the same time have a relative code redundancy equal to 0.44 (44%), which allows the designing of *Fibonacci microprocessors for noise-resistant calculations.* The error-detection ability in such microprocessors exceeds 99% and with increase of bits can reach 99.9% and even higher. After the publication of the books [6, 19], the interest in these ideas in the world computer community has increased significantly. In this regard, for example, a counting device, based on the Fibonacci code, designed in 2012 by the Ukrainian scientists Alexey Borisenko and Alexey Stakhov, has attracted great technical and scientific interest. This counting device is currently patented in Ukraine under the name "Noise-tolerant impulse counter by Borisenko–Stakhov". In essence, this technical device can be the beginning of a new element base for noise-tolerant microprocessors and microcontrollers for mission-critical applications and opens up new paths in the development of specialized computer technology. There are surprising analogies between the *Fibonacci code* and the *genetic code*, which single out the *Fibonacci code* (and the *code of the golden ratio*) as a special type of positional numeral systems, in which the natural principles of introduction of code redundancy into informational systems are implemented. **These new numeral systems are a prerequisite for the "golden" computer revolution in the specialized computer technology.**

(12) **The 12th conclusion** concerns the connection between *Mathematics of Harmony* and modern education. *Mathematics of Harmony* is an interdisciplinary discipline, which contains all the necessary components to become a link for all disciplines, which form the basis of modern education: *mathematics, physics, chemistry, botany, biology, medicine, geology, computer science, economy,* and *fine and applied art*. At the same time, the study of mathematics, which, as a rule, does not arise much interest in the majority of students and seems to be a *dry* and *boring* discipline, turns into an exciting search for the mathematical laws of Nature. The introduction of the ideas of *Harmony* and the *golden section* in modern education is an urgent problem because it can lead to the formation of the students' *new scientific worldview* based on the ancient ideas of *Harmony* and the *golden section*. During the period 2000–2001, the author delivered the course *Mathematics of Harmony and the golden section* for the students of *mathematics, physics,* and *computer science* of the Vinnitsa Pedagogical University. The program of the educational discipline *Mathematics of Harmony and the golden section* has been published in the author's article [155].

(13) **The 13th conclusion** concerns the role, which the *Mathematics of Harmony* can play in overcoming the *strategic mistakes* in the contemporary crisis in mathematics. Perhaps, the most in-depth and professional discussion of this crisis is given in the article "Pseudoscience: a disease that there is no one to cure" [128], published by Denis Kleschev on the website of the Academy of Trinitarism. The same problem is discussed in Stakhov's article [129], published on the website of the Academy of Trinitarism. These include the following: *neglect of beginnings, disregard for the idea of harmony and the golden section, ignoring Proclus's hypothesis*, concerning the influence of the harmonic ideas of Pythagoras and Plato on the creation of the Euclidean *Elements* and the origin of mathematics, *underestimation of Binet's formulas, a new class of hyperbolic functions based on the golden proportion*, the *neglect of Felix Klein's icosahedral idea*, and finally the *underestimation of the Bergman numeral system*, which turns our views on the numeral systems. However, the big strategic mistake of the 19th century mathematicians was a *revaluation of the Cantor theory of infinite sets*, despite the negative assessments of this theory by some famous of Cantor contemporaries (Kronecker, Poincaré, and others). Unfortunately, the prognosis of Henri Poincaré that *the future generations will see set theory as a disease, from which they must be cured* came true. The final chapter of Morris Klein' book [101] is called *The Authority of Nature*. In this chapter, Klein refers to the opinions of the outstanding scientists, who emphasize the role of Nature and theoretical natural sciences in the development of mathematics. In particular, one of the outstanding specialists in the foundations of mathematics, the Poland mathematician Andrzej Mostowski, states that *"mathematics belongs to the natural sciences. Mostowski's appeal to experience, and any attempt to substantiate mathematics without the appeal to its natural basis and applications, or even neglecting by history of mathematics is doomed to failure"*. Hermann Weil, who *"openly advocates considering mathematics as one of the natural sciences,"* adheres to the same point of view. But after all, *Mathematics of Harmony* is a mathematical discipline, which, in essence, is some kind of *Mathematics of Nature*. The main

goal of *Mathematics of Harmony* is a search for new mathematical constants, recurrent relations, and algebraic equations, which can be used to *model harmonic structures and natural phenomena*. This is why the rapprochement of *Mathematics of Harmony* with *Classical Mathematics* will help both to overcome the *strategic mistakes* in the development of mathematics and bring it closer to theoretical natural sciences.

(14) **The 14th conclusion** concerns the answer to the question: *Is Mathematics of Harmony the "golden" paradigm of modern science?* The essence of the "golden" paradigm of ancient Greek science is most clearly formulated in the well-known statement of Alexey Losev [5]: *"From Plato's point of view, and in general from the point of view of all ancient cosmology, the world is a proportional whole, obeying to the law of harmonic division, the golden section."* The interest in the "golden" paradigm of the ancient Greeks increased during the heydays of human culture, in particular, in the ancient Greek era and the Renaissance. But the highest surge of interest in the "golden" paradigm is observed in the second half of the 20th century and the beginning of the 21st century, when the professional groups of scientists were created to study this problem. The most famous of these are the *American Fibonacci Association* (1963), the *Slavic Golden Group* (1992), and the *International Club of the Golden Section* (2003). During this period, different international forums were held on this topic. The most famous of them are the *International Conferences on Fibonacci numbers and their applications*, which are held regularly (once every 2 years) by the *Fibonacci Association* in various countries since 1984. From 1992 to 1996 in Kiev (1992, 1993), and then in Stavropol (1994–1996), the International Seminars were held on the theme *"Golden Proportion and Problems of Harmony of Systems"*. In 2003, in the Ukrainian city of Vinnitsa, the International Conference *"Problems of Harmony, Symmetry and the Golden Section in Nature, Science and Art"* was held, and in 2010, the *International Congress on Mathematics of Harmony* was held in the Ukrainian city of Odessa. In recent years,

three unique books, related to the *Mathematics of Harmony* and the *golden ratio*, were published:

(1) **Stakhov Alexey. Assisted by Scott Olsen.** *The Mathematics of Harmony. From Euclid to Contemporary Mathematics and Computer Science.* World Scientific, 2009.
(2) *The Prince of Wales. Harmony. A New Way of Looking at our World.* An Imprint of Harper Collins Publisher, 2010.
(3) **Hrant Arakelyan.** *Mathematics and History of the Golden Section.* Moscow: Publishing House "Logos", 2014.

All these facts testify to the great interest of the modern scientific community to this problem. Currently, there is the process of *Harmonization of modern theoretical natural sciences*, which is accompanied by the process of *Harmonization of modern mathematics* [68]. Therefore, the creation of a new interdisciplinary direction, called *Mathematics of Harmony* [6], became a natural and logical reflection of one of the most important trends of modern science, the *revival of the "golden" paradigm of ancient Greeks in modern science*, which was embodied by Euclid in his *Elements*, the greatest mathematical work of ancient times.

(15) **The 15th conclusion concerns** the mystery of *Harmony of the Universe* as one of the greatest mysteries of Nature. Modern science is unable to answer not only the questions about the mystery of *the origin of the World*, the mystery of *birth and death*, the mystery of the *origin of Human Mind*, but also the questions about the *mystery of the genetic code* and the *riddle of the botanical phenomenon of phyllotaxis*, which have highly harmonic and symmetrical properties and which are studied actively in modern "materialistic science". The prominent Russian scientist, Doctor of Physical and Mathematical Sciences and PhD in Biological Sciences, Sergey Petoukhov, who devoted many years to research in the field of genetic coding and published many works in this field, came to the following conclusion [48]:

"Modern science is not aware of the reasons why the alphabet of the genetic language is four-letter (and not of thirty letters, for example); why out of the billions of possible chemical compounds, these four nitrogen bases of C, A, G, U (T) are chosen as elements of the alphabet; why exactly 20 amino acids are genetically encoded, etc."

To answer this question and other questions, posed by Petoukhov, in particular, the question about the surprisingly symmetrical nature of the unique octet matrix of the genetic code, we inevitably refer to the idea of the "Creator" or the "Universal Mind". We come to the same conclusion when analyzing the *new geometric theory of phyllotaxis* created by the talented Ukrainian researcher Doctor of Art History Oleg Bodnar [28]. Bodnar's Geometry has raised many new questions in modern science, in particular the following question. If Nature actually acts according to the scenario, proposed by Oleg Bodnar [28], then we must recognize that *Nature is an excellent mathematician*; Nature has known and used in its objects the hyperbolic Fibonacci and Lucas functions for millions or even billions of years. However, the materialistic approach to this problem and the history of the development of science tells us that all unsolved scientific problems will be solved sooner or later. *Mathematics of Harmony*, developed in the works of the author and other authors, is a source of new and original solutions in the field of modern mathematics, computer science and theoretical natural sciences [1–4, 6–11, 16, 17, 19, 20, 23, 24, 28, 30–35, 37, 38, 40–100, 119, 127–129, 136, 141–145, 151, 152, 154–156, 161, 162].

Bibliography

The Books in the Field of the Golden Section, Fibonacci Numbers and Mathematics of Harmony

[1]   Shevelev I.Sh., *Meta-Language of Wildlife*. Moscow: Sunday (2000) (in Russian).

[2]   Shevelev I.Sh., *The Principle of Proportion*. Moscow: Stroiizdat (1986) (in Russian).

[3]   Shevelev I.Sh., Marutaev M.A., Shmelev I.P., *Golden Section. Three Views on the Harmony of Nature*. Moscow: Stroiizdat (1990) (in Russian).

[4]   Soroko E.M., *Structural Harmony of Systems*. Minsk: Science and Technology (1984) (in Russian).

[5]   Losev A., *The History of Philosophy as a School of Thought*. Communist **11** (1981) (in Russian).

[6]   Stakhov A.P., *The Mathematics of Harmony. From Euclid to Contemporary Mathematics and Computer Science*. Assisted by Scott Olsen. World Scientific (2009).

[7]   Coxeter H.S.M., *Introduction to Geometry*. New York: John Wiley & Sons (1961).

[8]   Vorobyov N.N., *Fibonacci Numbers*. Moscow: Science (1984) (first edition, 1961) (in Russian).

[9]   Hoggatt V.E. Jr., *Fibonacci and Lucas Numbers*. Boston, MA: Houghton Mifflin (1969).

[10] Shestakov V.P., *Harmony as an Aesthetic Category*. Moscow: Science (1973) (in Russian).

[11] Vajda S., *Fibonacci & Lucas Numbers, and the Golden Section. Theory and Applications*. Ellis Harwood Limited (1989).

[12] Martin G., *Mathematics, Magic and Mystery*. New York: Dover Publications (1952).

[13] Brousseau A., *An Introduction to Fibonacci Discovery*. San Jose, CA: Fibonacci Association (1965).

[14] Huntley H.E., *The Divine Proportion: A Study in Mathematical Beauty*. Dover Publications (1970).

[15] Ghyka M., *The Geometry of Art and Life*. Dover Publications (1977).

[16] Stakhov A.P., *Introduction to Algorithmic Measurement Theory*. Moscow: Soviet Radio (1977) (in Russian).

[17] Stakhov A.P., *Algorithmic Measurement Theory*. Moscow: Knowledge (1979). (New in Life, Science and Technology. " Series Mathematics and Cybernetics", No. 6) (in Russian).

[18] Rigny A., *The Trilogy of Mathematics*. Translated from Hungarian. Moscow: World (1980) (in Russian).

[19] Stakhov A.P., *Codes of the Golden Proportion*. Moscow: Radio and Communications (1984) (in Russian).

[20] Grzedzielski J., *Energetycno-geometryczny kod Przyrody*. Warszawa: Warszwskie centrum studenckiego ruchu naukowego (1986) (in Polen).

[21] Garland T.H., *Fascinating Fibonacci: Mystery and Magic in Numbers*. Dale Seymour (1987).

[22] Kovalev F., *The Golden Ratio in Painting*. Kiev: High School (1989) (in Russian).

[23] Noise-immune codes, *Fibonacci Computer*. Series of Radio Electronics and Telecommunications, No. 6. Moscow: Knowledge (1989) (in Russian).

[24] Vasyutinsky N.A., *Golden Proportion*. Moscow: Young Guard (1990) (in Russian).

[25] Runion G.E., *The Golden Section*. Dale Seymour (1990) (in Russian).

[26] Robert F., *Fibonacci Applications and Strategies for Traders*. New York: John Wiley & Sons (1993).

[27] Shmelev I.P., *The Phenomenon of Ancient Egypt*. Minsk: RITS (1993) (in Russian).

[28] Bodnar O.Ya., *The Golden Section and Non-Euclidean Geometry in Nature and Art*. Lvov: Sweet (1994) (in Russian).

[29] Dunlap R.A., *The Golden Ratio and Fibonacci Numbers*. World Scientific (1997).

[30] Tsvetkov V.D., *Heart, Golden Proportion and Symmetry*. Pushchino (1997) (in Russian).

[31] Korobko V.I., *The Golden Proportion and the Problems of Harmony of Systems*. Moscow: Publishing House of the Association of Building Universities of the CIS countries (1998) (in Russian).

[32] Herz-Fischler R., *A Mathematical History of the Golden Number.* New York: Dover Publications (1998).

[33] de Spinadel V.W., *From the Golden Mean to Chaos.* Nueva Libreria (1998) (second edition, Nobuko, 2004).

[34] Gazale M.J., *Gnomon. From Pharaohs to Fractals.* Princeton, NJ: Princeton University Press (1999).

[35] Prechter R.R., *The Wave Principle of Human Social Behaviour and the New Science of Socionomics.* Gainesville, GA: New Classics Library (1999).

[36] Koshy T. *Fibonacci and Lucas Numbers with Applications.* New York: Wiley (2001).

[37] Kappraff J., *Connections. The Geometric Bridge Between Art and Science.* Second Edition. Singapore: World Scientific (2001).

[38] Kappraff J., *Beyond Measure. A Guided Tour Through Nature, Myth and Number.* Singapore: World Scientific (2002).

[39] Livio M., *The Golden Ratio: The Story of Phi, the World's Most Astonishing Number.* New York: Broadway Books (2002).

[40] Stakhov A.P., *New Math for Wildlife. Hyperbolic Fibonacci and Lucas Functions.* Vinnitsa: ITI (2003) (in Russian).

[41] Stakhov A.P., *Under the Sign of the "Golden Section". Confession of the Son of Studbat's Soljer.* Vinnitsa: ITI (2003) (in Russian).

[42] Bodnar O.Ya., *The Golden Section and Non-Euclidean Geometry in Science and Art.* Lviv: Ukrainian Technologies (2005) (in Ukrainian).

[43] Petrunenko V.V., *The Golden Section of Quantum States and Its Astronomical and Physical Manifestations.* Minsk: Law and Economics (2005) (in Russian).

[44] Dimitrov V., *A New Kind of Social Science. Study of Self-Organization of Human Dynamics.* Morrisville Lulu Press (2005).

[45] Soroko E.M., *The Golden Section, the Processes of Self-Organization and the Evolution of Systems. Introduction to the General Theory of System's Harmony.* Moscow: URSS (2006) (in Russian).

[46] Stakhov A., Sluchenkova A., Shcherbakov I., *Da Vinci Code and Fibonacci Numbers.* St. Petersburg: Peter (2006) (in Russian).

[47] Olsen S., *The Golden Section: Nature's Greatest Secret.* New York: Walker Publishing Company (2006).

[48] Petoukhov S.V., *Matrix Genetics, Algebras of Genetic Code, Noise Immunity.* Moscow-Izhevsk: Research Center "Regular and Chaotic Dynamics" (2008) (in Russian).

[49] The Prince of Wales with Tony Juniper and Ian Scelly, *Harmony: A New Way of Looking at Our World.* New York: HarperCollins Publisher (2010).

[50] Arakelian H., *Mathematics and History of the Golden Section.* Moscow: Logos (2014) (in Russian).

[51] Stakhov A., Aranson S., *The Mathematics of Harmony and Hilbert's Fourth Problem. The Way to Harmonic Hyperbolic and Spherical Worlds of Nature.* Germany: Lambert Academic Publishing (2014).

[52] Stakhov A., Aranson S., Assisted by Scott Olsen. *The "Golden" Non-Euclidean Geometry,* World Scientific (2016).

[53] Stakhov A., *Numeral Systems with Irrational Bases for Mission-Critical Applications.* World Scientific (2017).

The Articles in the Field of the Golden Section, Fibonacci Numbers and Mathematics of Harmony

[54] Bergman G., A number system with an irrational base. *Mathematics Magazine* **31**, (1957).

[55] Stakhov A.P., Redundant binary positional numeral systems. In *Homogenous Digital Computer and Integrated Structures*, No. 2. Taganrog: Publishing House "Taganrog Radio University" (1974) (in Russian).

[56] Stakhov A.P., An use of natural redundancy of the Fibonacci number systems for computer systems control. *Automation and Computer Systems* **6**, (1975) (in Russian).

[57] Stakhov A.P., Principle of measurement asymmetry. *Problems of Information Transmission* **3**, (1976) (in Russian).

[58] Stakhov A.P., Digital metrology in the fibonacci codes and the golden proportion codes. In *Contemporary Problems of Metrology.* Moscow: Publishing House of Moscow Machine-Building Institute (1978) (in Russian).

[59] Stakhov A.P., The golden mean in digital technology. *Automation and Computer Systems* **1**, (1980) (in Russian).

[60] Stakhov A.P., Algorithmic measurement theory and fundamentals of computer arithmetic. *Measurement, Control, Automation* **2**, (1988) (in Russian).

[61] Stakhov A.P., The golden section in the measurement theory. *Computers & Mathematics with Applications* **17**(4–6), (1989).

[62] Stakhov A.P., The golden proportion principle: Perspective way of computer progress. *Bulletin of the Academy of Sciences of Ukraine* **1–2**, (1990) (in Ukrainian).

[63] Stakhov A.P., The golden section and science of system harmony. *Bulletin of the Academy of Sciences of Ukraine* **12**, (1991) (in Ukrainian).

[64] Stakhov A.P., Tkachenko I.S., Hyperbolic Fibonacci trigonometry. *Reports of the National Academy of Sciences of Ukraine* **208**(7), (1993) (in Russian).

[65] Stakhov A.P., Algorithmic measurement theory: A general approach to number systems and computer arithmetic. *Control Systems and Computers* **4–5**, (1994) (in Russian).

[66] Stakhov A.P., The golden section and modern harmony mathematics. *Applications of Fibonacci Numbers* **7**, (1998).

[67] Spears C.P., Bicknell-Johnson M., Asymmetric cell division: Binomial identities for age analysis of mortal vs. immortal trees. *Applications of Fibonacci Numbers* **7**, (1998).

[68] Stakhov A.P., *Mathematization of Harmony and Harmonization of Mathematics.* Moscow: Academy of Trinitarism, El. No. 77-6567, publ. 166897 (2011) (in Russian).

[69] Stakhov A.P., A generalization of the Fibonacci Q-matrix. *Reports of the National Academy of Sciences of Ukraine* **9**, (1999).

[70] Stakhov A., Matrix arithmetics based on Fibonacci matrices. *Samara-Moskow: Computer Optics* **21**, (2001).

[71] Stakhov A., Ternary mirror-symmetrical arithmetic and its applications to digital signal processing. *Samara-Moskow: Computer Optics* **21**, (2001).

[72] Stakhov A.P., Brousentsov's ternary principle, Bergman's number system and ternary mirror-symmetrical arithmetic. *The Computer Journal* **45**(2), (2002).

[73] Radyuk M.S., The second golden section (1,465 ...) in Nature. Proceedings of the international conference "Problems of harmony, symmetry and the golden section in nature, science and art. Vinnitsa State Agrarian University, No. 15 (2003) (in Russian).

[74] Stakhov A.P., Generalized golden sections and a new approach to the geometric definition of a number. *Ukrainian Mathematical Journal* **56**(8), (2004) (in Russian).

[75] Stakhov A., Rozin B., On a new class of hyperbolic function. *Chaos, Solitons & Fractals* **23**(2), (2005).

[76] Stakhov A.P., The generalized principle of the golden section and its applications in mathematics, science, and engineering. *Chaos, Solitons & Fractals* **26**(2), (2005).

[77] Stakhov A., Rozin B., The golden shofar. *Chaos, Solitons & Fractals* **26**(3), (2005).

[78] Stakhov A.P., Golden section, sacred geometry and mathematics of harmony. In *Metaphysics. Century XXI.* Collection of Papers. Moscow: BINOM (2006) (in Russian).

[79] Stakhov A.P., Fundamentals of the new kind of Mathematics based on the golden section. *Chaos, Solitons & Fractals* **27**(5), (2006).

[80] Stakhov A., Rozin B., The continuous functions for the Fibonacci and Lucas *p*-numbers. *Chaos, Solitons & Fractals* **28**(4), (2006).

[81] Stakhov A., Fibonacci matrices, a generalization of the "Cassini formula," and a new coding theory. *Chaos, Solitons & Fractals* **30**(1), (2006).

[82] Stakhov A.P., Gazale formulas, a new class of the hyperbolic Fibonacci and Lucas functions, and the improved method of the "golden" cryptography. Moscow: Academy of Trinitarism, No. 77-6567, publication 14098 (2006) (in Russian).

[83] Stakhov A., The "golden" matrices and a new kind of cryptography. *Chaos, Solitons & Fractals* **32**(3), (2007).

[84] Stakhov A.P., The generalized golden proportions, a new theory of real numbers, and ternary mirror-symmetrical arithmetic. *Chaos, Solitons & Fractals* **33**(2), (2007).

[85] Stakhov A.P., Three "key" problems of mathematics on the stage of its origin and new directions in the development of mathematics, theoretical physics and computer science. Moscow: Academy of Trinitarism, No. 77-6567, publication 14135 (2007) (in Russian).

[86] Stakhov A.P., The mathematics of harmony: Clarifying the origins and development of mathematics. *Congressus Numerantium* **193**, (2008).

[87] Stakhov A.P., Aranson S.Ch., "Golden" Fibonacci goniometry, Fibonacci–Lorentz transformations, and Hilbert's fourth problem. *Congressus Numerantium* **193**, (2008).

[88] Stakhov A.P., Aranson S.Ch., Hyperbolic Fibonacci and Lucas functions, "golden" Fibonacci goniometry, Bodnar's geometry, and Hilbert's fourth problem. Part I. Hyperbolic Fibonacci and Lucas functions and "Golden" Fibonacci goniometry. *Applied Mathematics* **2**, (2011).

[89] Stakhov A.P., Aranson S.Ch., Hyperbolic Fibonacci and Lucas functions, "golden" Fibonacci goniometry, Bodnar's geometry, and Hilbert's fourth problem. Part II. A new geometric theory of phyllotaxis (Bodnar's geometry). *Applied Mathematics* **3**, (2011).

[90] Stakhov A.P., Aranson S.Ch., Hyperbolic Fibonacci and Lucas functions, "golden" Fibonacci goniometry, Bodnar's geometry, and Hilbert's fourth problem. Part III. An original solution of Hilbert's fourth problem. *Applied Mathematics* **4**, (2011).

[91] Stakhov A.P., Hilbert's fourth problem: Searching for harmonic hyperbolic worlds of nature. *Applied Mathematics and Physics* **1**(3), (2013).

[92] Stakhov A., A History, the main mathematical results and applications for the mathematics of harmony. *Applied Mathematics* **5**, (2014).

[93] Stakhov A., The mathematics of harmony. Proclus' hypothesis and new view on Euclid's elements and history of mathematics starting since Euclid. *Applied Mathematics* **5**, (2014).

[94] Stakhov A., The "golden" number theory and new properties of natural numbers. *British Journal of Mathematics & Computer Science* **6**, (2015).

[95] Stakhov A., Proclus hypothesis. *British Journal of Mathematics & Computer Science* **6**, (2016).

[96] Stakhov A., Aranson S., Hilbert's fourth problem as a possible candidate on the MILLENNIUM PROBLEM in geometry. *British Journal of Mathematics & Computer Science* **4**, (2016).

[97] Stakhov A., Fibonacci *p*-codes and codes of the golden *p*-proportions: New informational and arithmetical foundations of computer science and digital metrology for mission-critical applications. *British Journal of Mathematics & Computer Science* **1**, (2016).

[98] Stakhov A., Aranson S., The fine-structure constant as the physical-mathematical MILLENNIUM PROBLEM. *Physical Science International Journal* **1**, (2016).

[99] Stakhov A., The importance of the golden number for mathematics and computer science: Exploration of the Bergman's system and the Stakhov's ternary mirror-symmetrical system (numeral systems with irrational bases). *British Journal of Mathematics & Computer Science* **3**, (2016).

[100] Stakhov A., Mission-critical systems, paradox of hamming code, row hammer effect, 'Trojan Horse' of the binary system and numeral systems with irrational bases. *The Computer Journal* **61**(7), (2018).

Other Publications

[101] Klein M., *Mathematics. Loss of Certainty.* Translated from English. Moscow: World (1984) (in Russian).

[102] Kolmogorov A.N., *Mathematics in Its Historical Development.* Moscow: Science (1991) (in Russian).

[103] Harmony of spheres. *The Oxford Dictionary of Philosophy*, Oxford University Press (1994, 1996, 2005).

[104] *The Elements of Euclid. Books I–VI.* Translation from Greek and comments by DD Mordukhay-Boltovsky. Moscow-Leningrad (1948) (in Russian).

[105] *The Elements of Euclid. Books VII–X.* Translation from Greek and comments by DD Mordukhay-Boltovsky. Moscow-Leningrad (1949) (in Russian).

[106] *The Elements of Euclid. Books XI–XV.* Translation from Greek and comments by DD Mordukhay-Boltovsky. Moscow-Leningrad (1950) (in Russian).

[107] Khinchin A.Ya., *Chain Fractions.* Moscow: Fizmatgiz (1961) (first edition, 1935) (in Russian).

[108] Radoslav J., Pythagoras theorem and Fibonacci numbers http:// milan.milanovic.org/math/english/Pythagoras/Pythagoras.html.

[109] Korneev A.A., *Structural Secrets of the Golden Series.* Moscow: Academy of Trinitarism, El No. 77-6567, publ. 14359 (2007) (in Russian).

[110] Vilenkin N.V., *Combinatorics.* Moscow: Science (1969) (in Russian).

[111] Poya D., *Mathematical Discovery.* Translated from English. Moscow: Science (1970) (in Russian).

[112] Venninger M., *Models of Polyhedra.* Translation from English. Moscow: World (1974) (in Russian).

[113] Klein F., *Lectures on the Icosahedron and Solving Fifth-Degree Equations.* Moscow: Science (1989) (in Russian).

[114] Katz E.A., *Art and Science: About Polyhedra in General and the Truncated Icosahedron in Particular.* Moscow: Energy No. 10–12 (2002) (in Russian).

[115] Gratia D., Quasi-crystals. *Advances in the Physical Sciences,* **156**(2), (1988) (in Russian).

[116] Eletsky A.V., Smirnov B.M., Fullerenes. *Advances in the Physical Sciences* **163**(2), (1993) (in Russian).

[117] Kann Charles H., *Pythagoras and Pythagoreans. A Brief History.* Hackett Publishing Co, Inc. (2001).

[118] Vladimirov Yu.S., *Metaphysics.* Moscow: BINOM, Laboratory of Knowledge (2002) (in Russian).

[119] Vladimirov Yu.S., Quark icosahedron, charges and Weinberg angle. In *Proceedings of the International Conference "Problems of Harmony, Symmetry and the Golden Section in Nature, Science and Art"*, Vinnitsa (2003) (in Russian).

[120] Verkhovsky L.I., Platonic solids and elementary particles. *Chemistry and Life* **6**, (2006) (in Russian).

[121] Zhmud L., *The Origin of the History of Science in Classical Antiquity.* Walter de Gruyter (2006).

[122] Smorinsky C., *History of Mathematics. A Supplement.* Springer (2008).

[123] Donald K., *The Art of Computer Programming (TAOCP) in 4th Volumes*, Addison-Wesley (1962, 1968, 1969, 1973, 2005).

[124] Markov A.A., *On the Logic of Constructive Mathematics*. Moscow: Knowledge (1972) (in Russian).

[125] Hilbert D., On the infinite. In *Foundations of Geometry* (1948).

[126] Zenkin A.A., Error of George Cantor. *Problems of Philosophy* **2**, (2000) (in Russian).

[127] Stakhov A.P., Kleshchev D.S., *The Problem of the Infinite in Mathematics and Philosophy from Aristotle to Zenkin*. Moscow: Academy of Trinitarism, El. No. 77-6567, publ. 15680 (2009) (in Russian).

[128] Kleschev D., Pseudoscience: A Disease that There is No One to Cure. Moscow: Academy of Trinitarism, El. No. 77-6567, publ. 17012, (2011) (in Russian).

[129] Stakhov A.P., Is modern mathematics not standing on the "pseudoscientific" foundation? (Discussion of the article by Denis Kleshchev "Pseudoscience: a disease that there is no one to cure"). Moscow: Academy of Trinitariasm, El. No. 77-6567, publ. 17034 (2011) (in Russian).

[130] Weyl G., *On the Philosophy of Mathematics*. Moscow-Leningrad (1934), Reprint Moscow: KomKniga (2005) (in Russian).

[131] Wilde Duglas J., *Optimum Seeking Methods*. Translation from English. Moscow: Science (1967).

[132] Niccolò F.T., From Wikipedia, the free encyclopedia https://en. wikipedia.org/wiki/Niccol%C3%B2_Fontana_Tartaglia.

[133] Bashmakova I.G., Yushkevich A.P., The origin of the numeral systems. In the book "Encyclopaedia of elementary arithmetic". *Arithmetic*. Moscow-Leningrad: State Publishing House of Technical and Theoretical Literature (1951).

[134] Neugebauer O., *Lectures on the History of Ancient Mathematical Sciences. Volume 1 "Pre-Greek Mathematics"*. Translated from German. Moscow-Leningrad: United Scientific and Technical Publishing House of the USSR (1937) (in Russian).

[135] Pospelov D.A., *Arithmetic Foundations of Computers of Discrete Action*. Moscow: High School (1970) (in Russian).

[136] Borisenko A., *Binomial Counting. Theory and Practice*. Publishing House "University Book" (2004) (in Russian).

[137] Iliev L., Mathematics as a science about models. *Successes of Mathematical Sciences* **27**(2), (1972) (in Russian).

[138] Kartsev M.A., *Arithmetic of Digital Machines*. Moscow: Science (The main edition of the physical and mathematical literature) (1969) (in Russian).

[139]   Kuhn T.S., *The Structure of Scientific Revolutions*. Chicago: University of Chicago Press (1962) (Russian translation, 1975).

[140]   Shervatov V.G., *Hyperbolic Functions*. Moscow: Fizmatgiz (1958) (in Russian).

[141]   Tatarenko A.A., *Golden T_m-harmonies and D_m-fractals as the Essence of Soliton-like T_m-structure of the World*. Moscow: Academy of Trinitarism, El. 77-6567, publ. 12691 (2005) (in Russian).

[142]   *Metaphysics. Century XXI*. Collection of Works. Compiled and edited by Yury Vladimirov. Moscow: BINOM, Laboratory of Knowledge (2006).

[143]   Shenyagin V.P., *Pythagoras, or Everyone Creates Their Own Myth": Fourteen Years from the Moment of the First Publication on the Quadratic Mantissa s-proportions*. Moscow: Academy of Trinitarism, El. 77-6567, publ. 17031 (2011) (in Russian).

[144]   Kosinov N.V., *Golden Proportion, Golden Constants and Golden Theorems*. Moscow: Academy of Trinitarism, El. No. 77-6567, publ. 14379 (2007) (in Russian).

[145]   Falcon S., Angel P., On the Fibonacci k-numbers. *Chaos, Solitons & Fractals*, **32**(5), (2007) pp. 1615–1624.

[146]   Aleksandrov P.S., (General Editor), *Hilbert's Problems*. Moscow: (1969) (in Russian).

[147]   Demidov S.S., To the history of Hilbert problems. In *Historical and Mathematical Research*. Moscow: Science (1966), No. 17. pp. 91–122. (in Russian)

[148]   Demidov S.S., Hilbert Mathematical Problems and Mathematics of the XX century. In *Historical and Mathematical Research*. Moskow: Janus-K (2001), No. 41 (6). pp. 84–99 (in Russian)

[149]   Bolibruh A.A., *Hilbert Problems (100 years later)*. Moscow: Science (Publishing house of the Moscow center for continuous mathematical education) (1999), V. 2, 24 p. (Library "Mathematical education" (in Russian).

[150]   Matiyasevich Y.V., *Hilbert's 10-th Problem*. Moscow: Science (Mathematical logic and foundations of geometry) (1993) (in Russian).

[151]   Stakhov A.P., Aranson S.Kh., On insolvability of the 4-th Hilbert problem for hyperbolic geometries. *Journal of Advances in Mathematics and Computer Science* (2019) **31**(1), pp. 1–21; Article No. JAMCS. 47241. ISSN: 2456-9968 (Past name: *British Journal of Advances in Mathematics and Computer Science.* Past ISSN:2231-0851) (http://www.journaljamcs.com/index.php/JAMCS/article/view/30099.)

[152] Kleschev D., *About the Past and Future Gods, Priests and Prophets of Science*. Moscow: Academy of Trinitarism, El. No. 77-6567, publ. 16762 (2011) (in Russian).

[153] Klein, M., *Mathematics and the Search for Knowledge*, Oxford University Press (1985).

[154] Voloshinov A.V., *Mathematics and Art. First edition*, Moscow: Enlightenment (1992). Second edition, Moscow: Enlightenment (2000).

[155] Stakhov A.P., *The Program of the Course "Mathematics of Harmony and the Golden Section" for Physical and Mathematical Faculties of Pedagogical Universities*. Moscow: Academy of Trinitarism, El. 77-6567, publ. 12237 (2005) (in Russian).

[156] Abachiev, S., Mathematics of harmony through the eyes of the historian and expert of methodology of science, *Science about Science* **4** (2012).

[157] Aranson S., *Once Again on Hilbert's 4th Problem*. Moscow: Academy of Trinitarism, El. No. 77-6567, publ. 15677 (2009) (in Russian).

[158] Buseman G., On the 4th Hilbert problem. *Advances in Mathematical Sciences* **21**(1), (1966) pp. 155–164 (in Russian).

[159] Hamel, G., On the geometries, in which the straight lines are the shortest. *Mathematics Ann.* **67**, (1903) pp. 231–264 (in German).

[160] Pogorelov V., *The Hilbert's Fourth Problem*. Moscow: Science (1974) (in Russian).

[161] Arakelian H., *On World Harmony, The Theory of the Golden Ratio and Its Generalizations*. Moscow: Academy of Trinitarism, El. No. 77-6567, publ. 17064 (2011) (in Russian).

[162] Mitropolskiy Yu.A., *Feedback on the Scientific Direction of the Ukrainian Scientist, Doctor of Technical Sciences, Professor Aleksey Petrovich Stakhov*. Moscow: Academy of Trinitarism, El. No. 77-6567, publ. 12452, (2005) (in Russian).

SERIES ON KNOTS AND EVERYTHING

ISSN: 0219-9769

Editor-in-charge: Louis H. Kauffman *(Univ. of Illinois, Chicago)*

The Series on Knots and Everything: is a book series polarized around the theory of knots. Volume 1 in the series is Louis H Kauffman's Knots and Physics.

One purpose of this series is to continue the exploration of many of the themes indicated in Volume 1. These themes reach out beyond knot theory into physics, mathematics, logic, linguistics, philosophy, biology and practical experience. All of these outreaches have relations with knot theory when knot theory is regarded as a pivot or meeting place for apparently separate ideas. Knots act as such a pivotal place. We do not fully understand why this is so. The series represents stages in the exploration of this nexus.

Details of the titles in this series to date give a picture of the enterprise.

Published:

More information on this series can also be found at http://www.worldscientific.com/series/skae

CPSIA information can be obtained
at www.ICGtesting.com
Printed in the USA
LVHW022326030920
665021LV00003B/22

9 789811 213496